MISTEAKS

MISTEAKS

. . . and how to find them before the teacher does . . .

A Calculus Supplement

by Barry Cipra

Birkhäuser
Boston • Basel • Stuttgart

Barry A. Cipra
Department of Mathematics
Ohio State University
Columbus, Ohio 43229

Library of Congress Cataloging in Publication Data

Cipra, Barry A.
Misteaks [i.e. Mistakes] and how to find them
before the teacher does.

Bibliography: p.
1. Calculus. 2. Calculus—Problems,
exercises, etc. I. Title. II. Title:
Misteaks and how to find them before the
teacher does.
QA303.C578 1983 515 82-4553
ISBN 3-7643-3083-X AACR2

CIP—Kurztitelaufnahme der Deutschen Bibliothek

Cipra, Barry A.:
Misteaks [Mistakes] and how to find them before
the teacher does: a calculus suppl. / Barry
A. Cipra. — Boston; Basel; Stuttgart:
Birkhäuser, 1983.
ISBN 3-7643-3083-X

ISBN 3-7643-3083-X
Printed in USA

Contents

We think, therefore we err.

Introduction

Everybody makes mistakes. Young or old, smart or dumb, student or teacher, we all make 'em. The difference is, smart people try to *catch* their mistakes.

This book, by showing how mistakes can be looked for and found in calculus problems, can help make you look like a smart person—at least in calculus class.

Solving a mathematical problem consists of two essential steps: 1) writing down an answer; and 2) asking if that answer makes any sense. In theory, if the first step is carried out rigorously, by an error-free, computerlike mind, then the second step is redundant and unnecessary. In practice, however, both steps are necessary: mistakes both minor and major abound, and you can never be sure of their absence unless you check for their presence.

Unfortunately, only the first step ever seems to get taught. Why? It's not that teachers ascribe to their students any sort of perfection. (If you've ever sat around a faculty lounge, you'll know that's not the case!) And it's certainly not that teachers themselves are flawless and therefore don't recognize the need for checking their work.

If you thought I was rhetorically leading up to a grand explanation by diverting you with a couple of balancing misdirections (a standard pedagogical device), forget it. I don't know why the second step doesn't get taught. Maybe there's a secret society involved. Maybe the CIA is behind it. Who knows?

This book, then, is an attempt to divulge some of the secrets, if that's what they are, of the second step. The examples and techniques here are mostly related to the mistakes that occur in calculus problems. This does not necessarily restrict their significance. If you are attentive to the spirit of these methods, and ignore some of the sloppy exposition, you should find the ideas here generally applicable.

The subtitle for this book is perhaps itself in error. What I'm really trying to do here is show how to *look* for mistakes, not necessarily how to *find* them. As we'll see, there are lots of ways you can *know* that you've made a mistake without having any idea *where* you made it, or even exactly what kind of a mistake it was.

Of course, once you *know* you've made a mistake, then you're supposed to go *find* it, and ultimately you ought to *do* something about it. That can be difficult, especially in the limited time you have for an exam. And curiously, it always seems easier to find someone else's errors than it is to find your own. (In fact, a good way to read this book is to 'grade' it—look for where the mistakes are and think about why they might have been made, then figure out how much partial credit the thing is worth.) One explanation for this is the 'Fresh Eyes' theory: when you check over your own work, you tend to look at it the same way each time; someone else, by virtue of being a different person, will look at your work differently. He may also make mistakes—but they'll be different ones!

In fact, this is the underlying message of this book: to be a successful problem solver (and I don't mean just of mathematical problems), you have to try to think in many different ways—use many different pairs of eyes, so to speak. You'll still make mistakes; that's taken for granted. But if you approach a problem in enough different ways, if you think about it in enough different lights, if you look at it from all different angles (I could go on and on), whatever mistakes you made will eventually show up—and then you can get rid of them.

Finally, a word about typos. Nothing is more frustrating than a typographical error in a textbook. A diligent student can agonize for hours because he got a plus sign while the book got a minus sign—only to have the teacher finally say that the book just made a simple mistake. (He could at least call it a *stupid* mistake.) Here that should not be a problem; the formulas in this book were meant to be wrong. Nevertheless, it was important that some things be stated correctly, so in the interest of readability, this book has been subject to a painstakingly carful proofreading.

Note to Teachers

The remarks here are mostly meant for teachers. You students are welcome to read them if you want, but they're not a required part of the book. *This material will not be on the test.*

OK, teachers, I think we're alone now.

Some of you may be wondering what in the world do I think I'm doing. After all, mathematics is the science of righteousness and exactitude, is it not? If we say to students it's all right to make mistakes, won't they just take that as carte blanche (in a note to teachers I can use fancy words like that—and interrupt the train of thought, like this) now where was I? Oh yes, won't they just take that as carte blanche to do *all* their work wrong? And accuse us of being equally wrong when we try to correct them?

I doubt it. At least not while teachers have the last word at exam time. All I've tried to do is write an interesting book which may be helpful to students trying to learn calculus.

Besides, there is beneath it all a serious message to this book, and I hope now that only teachers are reading this. To really learn calculus—or any other subject, for that matter—you have to do more than solve a bunch of problems: you have to *think* about what you're doing. Too many students leave the thinking part up to the teacher. The student's job, it seems, is to plug stuff into equations and crank out 'answers,' while the teacher gets paid to grade them. (I can't imagine where they get this idea from, can you?) The problems they *'solve'* seem to be isolated from any recognizable reality; in this formal garden the question "Does this answer make any sense?" itself makes no sense. What this book tries to do, then, is make less formal our beautiful garden of mathematics, and to encourage students to enjoy themselves in it. If they attack it with hedge clippers every so often, so much the better. We can always use a little pruning.

Originally, this was supposed to be the section in which I was going to expound my educational philosophy and principles, but after three drafts of exceptionally bad writing which even I thought showed a poverty of ideas and a wealth of ignorance, I realized that I simply have no profound or original insights into the subject, at least none that are fit to print. I'm still suffering the aftereffects of the attempt: excessively long sentences, ponderous and inflated language (big words), forced metaphors, and dubious grammatical constructions (including the use of colons). Fortunately I waited until the end to write this most difficult section (I always had trouble communicating with teachers), so the rest of the book was not affected. It has its own problems, but not these. At any rate, I'm sorry I have nothing to say here, but that's just too bad.

Note to Students

This book is meant to be browsed through, rather than systematically studied. Although there is some logic to the arrangement of sections, it has little to do with the logical structure of calculus. In particular, integrals and derivatives are lumped together here, while most calculus courses treat first one and then the other. Therefore (a word you're going to see a lot of), if an example uses something that you haven't studied yet, skip the example. (The alternative is to beat your head against a wall trying to figure out what's going on. You and the wall should both try to avoid this.)

Although I've included exercises at the end of each section, thus making the book a true math book, the best exercise is to look for the errors in your own work (preferably before you turn it in to be graded). The mistakes here are for the most part ad hoc, invented by myself to make the points I wanted to make. Some of them therefore strain credibility, as in Who would do such a stupid thing? Eventually I would like to replace them with 'honest' errors, taken from homework or exams. So if you have any mistakes of which you are particularly proud, send them in. Your contributions will be much appreciated, by me and possibly by future readers.

Finally, if you read the Note to Teachers (and you weren't meant to), you may still be wondering what in the world I'm trying to do. Well, I'm not going to tell you either. All that stuff about trying to "get students to think, for a change" has some validity, but doesn't really mean much. It would be nice if this book were to have some long-range significance. In the meantime, let's hope it helps you get a few extra points on the next exam.

1. Integrals: The Power of Positive Thinking

If you remember nothing else from calculus, remember this: *A definite integral measures the area beneath a curve.* In particular, a *positive* function must have a *positive* integral. Thus

$$\int_{-2}^{1}(x^2+1)dx = \left(\frac{1}{3}x^3+x\right)\Big|_{-2}^{-1}$$

$$= \frac{-1}{3}-1-\frac{8}{3}-2 = -6$$

is clearly wrong, because x^2+1 is clearly positive. Similarly,

$$\int_0^{2\pi}\sqrt{1-sin^2\theta}\ d\theta = \int_0^{2\pi}\sqrt{cos^2\theta}\ \ d\theta$$

$$= \int_0^{2\pi}cos\theta\ d\theta = sin\theta\ \Big|_0^{2\pi} = 0-0 = 0$$

is also wrong, because the square-root function, by definition, always means the *positive* square root.

Of course not every function is positive, and you can't always just look at a function and immediately say if it's positive or not. (Is $x^2-2xy+y^2$ always non-negative? Is $x^2+3xy+y^2$? How about $3+4cos\theta+cos2\theta$?) But some problems *require* a positive answer. In particular, area problems should get a positive response, as should volume integrals. Don't *ever* settle for a negative answer to such a problem. Fudge, if you have to, but get rid of that minus sign! For instance, if you want to compute the area between the sine and cosine functions in the interval $0 \leq x \leq \pi$, but get

$$A = \int_0^{\pi}(cosx-sinx)dx = sinx+cosx\ \Big|_0^{\pi} = -1-1 = -2,$$

then the least you can do is take the absolute value of the

thing, and claim you got $A=2$. That's still the wrong answer (which is $2\sqrt{2}$), but at least you're in the ballpark.

Area is not the only thing that shouldn't be negative. Consider the following "word problem":

> The acceleration a bicyclist can apply decreases as he gets tired. Suppose the acceleration is given by $a(t)=27-t^2/100$ feet/second^2. Starting from rest at $t=0$, how far does the bicyclist go by the time he is again at rest?

Without explaining the steps, here is a 'solution':

$$a(t) = 27 - \frac{1}{100}t^2$$

$$v(t) = \int\left(27 - \frac{1}{100}t^2\right)dt = 27t - \frac{1}{300}t^3$$

$$s(t) = \int\left(27t - \frac{1}{300}t^3\right)dt = \frac{27}{2}t^2 - \frac{1}{75}t^4$$

$$v(t) = 0 \Rightarrow 27t = \frac{1}{300}t^3 \Rightarrow t = 90 \text{ seconds}$$

$$s(90) = \frac{27}{2}(90)^2 - \frac{1}{75}(90)^4 = -765450 \text{ feet}$$

So what's wrong? The sign, of course! That negative sign means that, somehow or other, the bicyclist, in spite of all his efforts, has managed to travel not forwards, but *backwards.* There must be something wrong, either with him or with us.

Here are some problems which can be checked in this manner. Decide for each if the final answer is sensible or not; if it is not, see if you can figure out where I went wrong. (Please be generous with the partial credit.)

1. $\int_0^{\pi/2} sin\theta d\theta = cos\theta \,\Big|_0^{\pi/2} = cos(\pi/2) - cos(0) = 0-1 = -1$

2. $\int_0^1 (x^2-1)dx = \frac{1}{3}x^3 - x \,\Big|_0^1 = \left(\frac{1}{3}-1\right) - (0-0) = -2/3$

3. $\int_{-1}^{-2} x^2 dx = \frac{1}{3}x^3 \,\Big|_{-1}^{-2} = (-1/3)-(-8/3) = 7/3$

4. $\int_0^{\pi/4} (sin\theta - cos\theta)d\theta = -cos\theta - sin\theta \,\Big|_0^{\pi/4}$

$$= \left(\frac{-\sqrt{2}}{2} - \frac{-\sqrt{2}}{2}\right) - (-1-0) = 1-\sqrt{2}$$

5. $\int_{-1}^{1} \frac{dx}{x^2+1} = log(x^2+1) \Big|_{-1}^{1} = log(2) - log(2) = 0$

6. $\int_{-2}^{-1} \frac{dx}{x^2} = \frac{-1}{x} \Big|_{-2}^{-1} = (-1/1) - (-1/2)$

$$= -1 + \frac{1}{2} = -1/2$$

7. $\int_{-2}^{1} \frac{dx}{x^2} = \frac{-1}{x} \Big|_{-2}^{1} = (-1/1) - (-1/-2)$

$$= -1 - \frac{1}{2} = -3/2$$

8. $\int_{0}^{\pi/2} e^{sinx} cosxdx = e^{cosx} \Big|_{0}^{\pi/2} = 1-e$

9. $\int_{-\infty}^{\infty} e^{-x^2} dx = \frac{-1}{2x} e^{-x^2} \Big|_{-\infty}^{\infty} = 0-0 = 0$

2. Differentiating Right From Wrong

What the last section said for integrals, this section says for derivatives. If you remember nothing else from calculus, remember this: *A derivative measures the slope of a tangent line.* In particular, an increasing function must have a positive

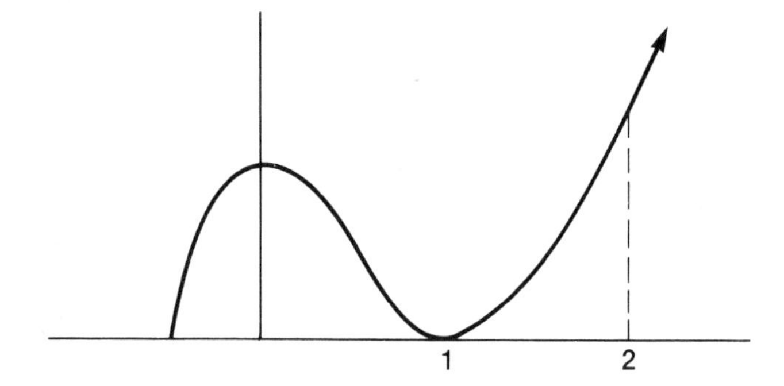

derivative. Thus for $f(x)=2x^3-3x^2+1$, the derivative $f'(x)=2x^2-6x$ gives the wrong answer at $x=2$: $f'(2)=2(2)^2-6(2)=8-12=-4$ is clearly wrong, because at $x=2$, the function is clearly *rising*. Trying $f'(x)=3x^2-6x$ is no better: $f'(2)=0$, but the function is still most definitely rising. Correcting this to $f'(x)=6x^2-6$ 'fixes' $x=2$, ($f'(2)=18$ is finally positive), but looks bad for $x=0$: $f'(0)=-6$, in spite of the fact that the function is *flat* there. (If you're getting tired of this nonsense, set $f'(x)=6x^2-6x$ and check that this gives reasonable values at $x=0,1,2$, and anywhere else you feel like checking.)

Of course, you may say, that's a cheat: I only knew $f'(2)$ should be positive because I had the picture to stare at. How many teachers are kind enough to provide pictures for every problem?

It's true you don't always (read: almost never) get a picture, but for many mistakes you don't need one. It's enough to have some general, vague idea what the function looks like. For instance, if $f(x) = xe^{-x}$, then $f'(x) = (x-1)e^{-x}$ is wrong for the following reason: $f'(0) = (0-1)e^{-0} = -1$ says the function is decreasing; since $f(0) = 0$, this means $f(x)$ would be *negative* for small values of x (you're going *down* from zero); but that's nonsense: $f(x) = xe^{-x}$ is clearly *positive* for *all* (positive) values of x.

Sometimes two well-chosen values can tell you what a function is doing. Consider

$$f(x) = \frac{x+5}{x+1} ,$$

and its 'derivative'

$$f'(x) = \frac{(x+5)-(x+1)}{(x+1)^2} = \frac{4}{(x+1)^2} .$$

Since this expression for the derivative is clearly always positive, we would expect the function $f(x)$ to always increase. But look what happens: $f(0) = 5/1 = 5$, while $f(1) = 6/2 = 3$. The values are going *down* instead of up.

Finally, the sign of a derivative is sometimes hinted at by the problem itself, especially if it's one of those nasty "word problems":

> A fungus culture grows until it fills its Petri dish, according to the law $F(t) = 1 - e^{-2t}$. Find the rate of growth when the dish is half full.

This problem has assured us (more or less) that the culture is *growing*. Thus the rate of growth should always be positive. For that reason, the 'answer' $F'(\frac{1}{2} log \frac{3}{2}) = -2e^{-log(3/2)} = -4/3$ *cannot* be right. There's only one thing to do: fudge. Put in *plus* signs, and claim $F'(\frac{1}{2} log \frac{3}{2}) = +4/3$.

Incidentally, doing this results in a 'double fudge', because I already fudged once in picking $t = \frac{1}{2} log(3/2)$ as the half-full time. Here's how:

You find t by setting $F(t) = 1/2$. Thus $\frac{1}{2} = 1 - e^{-2t}$, so that $e^{-2t} = 1 + \frac{1}{2} = \frac{3}{2}$. Thus $-2t = log(3/2)$, or $t = -\frac{1}{2} log(3/2)$.

But this is ridiculous: $-\frac{1}{2}log(3/2)$ is *negative*, and the Petri dish can't possibly be half full *before* it gets started! The value we want for t has to be positive, so there's only one thing to do: fudge—erase the minus sign and let $t = \frac{1}{2}log(3/2)$.

Please don't get the idea that fudging always gives the right answer. Sometimes it does and sometimes it doesn't. In the example above, one fudge did and one fudge didn't. (Unfortunately, the one that didn't preceded the one that did, so even the one that did really didn't.)*

What fudging does do is to turn an *obviously* wrong answer into something that *might* be correct. Of course what you 'should' do is "justify" your fudge—look back until you find a minus sign that really should have been a plus sign, and only then make corrections. That is, you really ought to find the *source* of your mistake before changing your answer. But on a one-hour test who has time for such luxuries? Fudge and forget it!

Here are some more problems with loads of mistakes, mostly of a negative nature (but a few which are *positively* wrong also):

1. Drawn below is the graph of a function $f(x)$, whose formula you don't need to know. Decide which of the expressions next to it are at all reasonable and which are impossible.

$f'(-2) = 2$
$f'(0) = 2$
$f'(1) = 3$
$f'(2) = 1$
$f(0)f'(0) = 1$
$f'(2)f(1) = 3$
$f'(2)f(2) = 2$
$f'(2)f(-2) = 1$

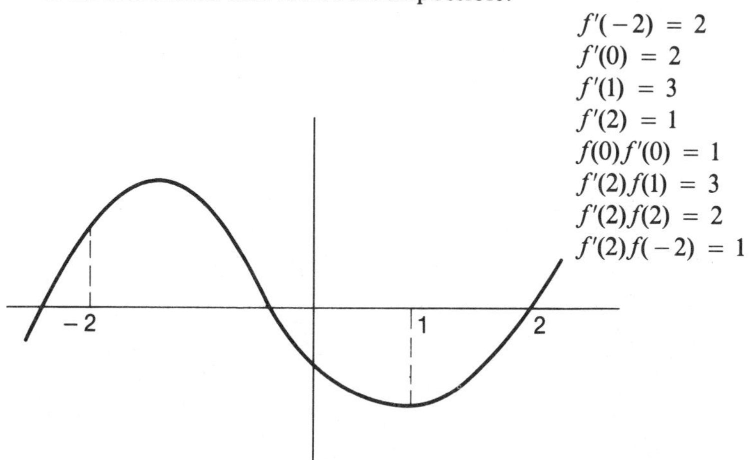

*I didn't understand that parenthetical remark either. (You can safely ignore all parenthetical remarks in this book.)

2. Find fault with these derivatives (if you can):
   a. $f(x) = x\cos^2 x \quad f'(0) = -1$
   b. $f(x) = (x-1)(x-2) \quad f'(2) = -2$
   c. $f(x) = \sqrt{1-\cos x} \quad f'(0) = -1/\sqrt{2}$
   d. $f(x) = \log(1-x) \quad f'(0) = -1 \quad f'(1) = 0$
   e. $f(x) = 1/x^2 \quad f'(x) = 2/x^3$
   f. $f(x) = \sin^3 x \quad f'(x) = 3\cos^2 x$ (Hint: the proposed derivative is always positive.)*
3. Discuss the positivity or negativity of the change in temperature of a bucket of water when the following items are dropped in it:
   a. an ice cube
   b. a glowing coal
   c. a used (or unused) calculus book
4. Consider the function $f(x) = (x-2)/(2x-1)$. Note that $f(0) = (-2)/(-1) = 2$, while $f(1) = (1-2)/(2-1) = -1$, implying that $f(x)$ *decreases* from 0 to 1. Nevertheless, the derivative is $f'(x) = 3/(2x-1)^2$ (check me!), which is obviously always *positive*. What's wrong here? Has this section been selling you a bill of goods?

*I can hear the teachers grinding their teeth: $\cos^2 x$ is *not* always positive. For instance, $\cos^2(\pi/2) = 0$. I was being sloppy; what I meant was, $f'(x)$ is always non-negative. This technicality has nothing to do with the hint.)

3. More Integrals (That's About The Size Of It)

We're not done yet with the technique of the previous sections. Remember: *A definite integral measures the area beneath a curve.* Perhaps a picture is in order:

$\int_a^b f(x)dx$ = area of the shaded region.

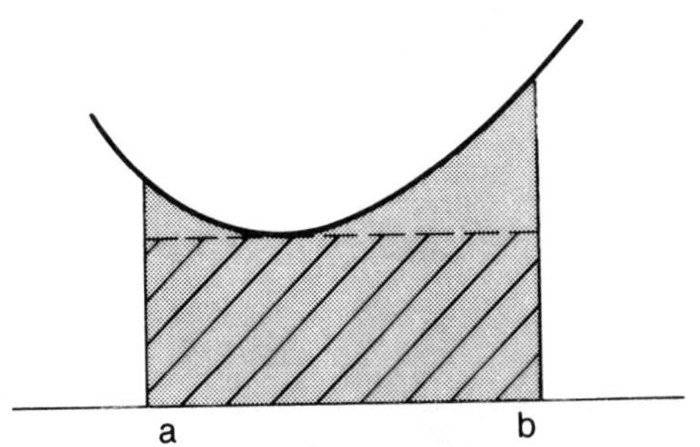

Now quite clearly, this dotted region contains a rectangle – the diagonally striped region. Obviously this rectangle must have smaller area. In the particular example we have drawn, $f(x)=x^2+1$, $a=-1/2$, and $b=1$, so the rectangle has base $b-a=3/2$, and height $f(0)=1$ (the lowest point of the parabola x^2+1), hence area $(3/2)(1)=3/2$. Thus

$$\int_{-1/2}^{1}(x^2+1)dx = \frac{1}{3}x^2+x\Big|_{-1/2}^{1}$$

$$=\left(\frac{1}{3}+1\right)-\left(\frac{1}{12}+\frac{1}{2}\right)$$

$$=\frac{16}{12}-\frac{7}{12}=\frac{3}{4}$$

is quite wrong, because 3/4 is *smaller* than 3/2, instead of larger, as it should be.

This cuts the other way, also: if the region is *contained* in some rectangle, then the integral must be *smaller* than the area of the rectangle:

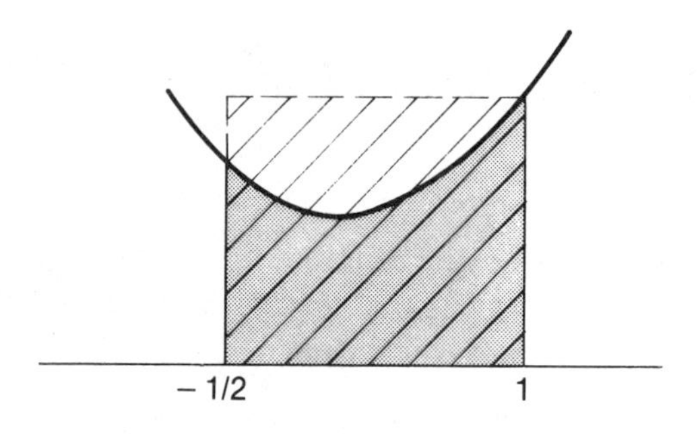

$$\int_{-1/2}^{1}(x^2+1)dx = (3x^2+x)\Big|_{-1/2}^{1}$$

$$= (3+1) - \left(\frac{3}{4} - \frac{1}{2}\right) = 15/4$$

is wrong, because the area of the outer rectangle is only $2(3/2) = 3$.

Rectangles aren't the only geometrical figures around. A few well-placed triangles can work wonders.

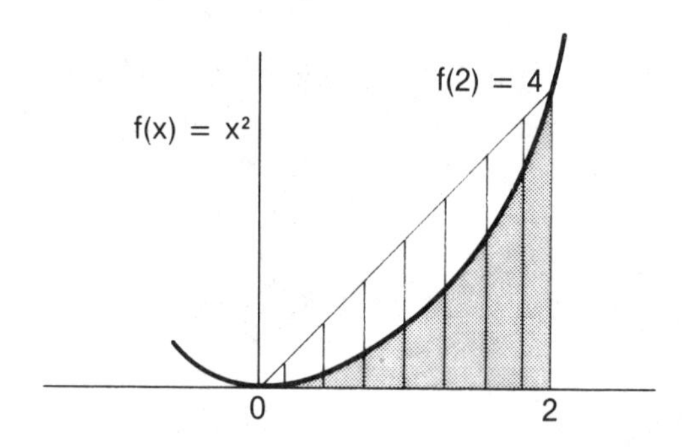

$$\int_0^2 x^2 dx = 1/2x^3 \Big|_0^2 = 8/2 - 0 = 4$$

is wrong, since the area of the triangle, which is $\frac{1}{2}bh = \frac{1}{2}(2)(4) = 4$ on the nose.

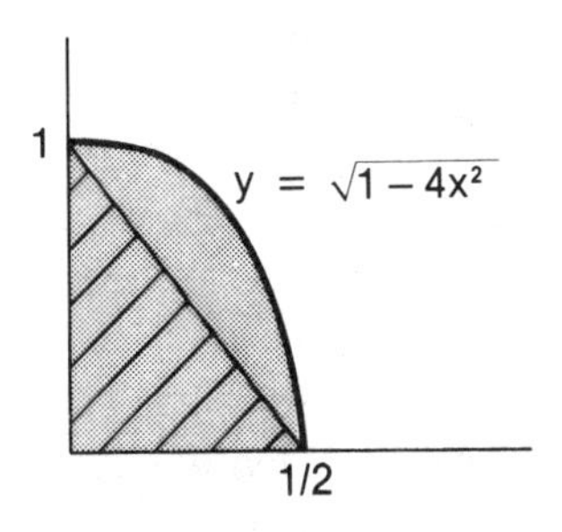

$$\int_0^{1/2}\sqrt{1-4x^2}\,dx = \int_0^{1/2}\sqrt{1-v^2}\,dv$$

$$= 1/2\int_0^{\pi/6}\sqrt{1-sin^2u}\,cosudu$$

$$= 1/2\int_0^{\pi/6}\left(\frac{1+cos2u}{2}\right)du = 1/4[u+ \tfrac{1}{2}sin2u]\ \Big|\ _0^{\pi/6}\mathrm{I}$$

$$= \pi/24+\sqrt{3}/16$$

is also wrong, though the check is subtle: the triangle has area $\frac{1}{2}(1/2)(1)=1/4$, while the 'answer' was $\pi/24+\sqrt{3}/16=(2\pi+3\sqrt{3})/48$. Now if the numerator is *smaller than 12*, we have found an error. Indeed it is: $2\pi+3\sqrt{3}=2(3.14159\ldots)+3(1.732051\ldots.)=11.47933\ldots.$. The triangle just barely found the error! (Actually, this example is something of a cheat: it required almost as much thought and work to find that $2\pi+3\sqrt{3}$ was less than 12 as it took to do the stupid problem in the first place.)

Notice something: in all these examples we used some 'geometric' knowledge about the function $f(x)$—the location of a minimum and maximum in the first two examples, and the concavity up or down in the last two. *It always helps to have some idea of what a function looks like.* A good picture makes any problem easier.

So how can you tell what a function looks like? How do you draw a picture?

Of course you learned in first semester calculus how to sketch a curve: You take the derivative and set it equal to zero. That gives you the maximums and the minimums (sometimes). Then you take the *second* derivative and set *it* equal to zero.

That gives you the inflection points (whatever they are). Where the first derivative is positive, the function increases; where negative, it decreases. When the second derivative is positive, the function is concave up (or is it down?), when negative, down (or is it up? and what is 'concavity' anyway? whatever happened to convexity?)

And what you probably discovered is that the whole process is a big mess. Taking one derivative is not so bad. Taking *two* derivatives is usually disastrous. And set that to zero and *solve*? Forget it!

The fact is, most of the functions you run into fall into one of two classes:

Class 1: the simple, everyday functions, whose graphs you already know (or should know). Examples are x^2, x^3, $\sqrt{x}$, e^x, e^{-x}, $logx$, $sinx$, $cosx$, $tanx$, $secx$. (If you can't give a quick, rough sketch of each of these, you'd better get on the stick.)

Class 2: functions so complicated, the methods you learned in first semester calculus won't do you a bit of good. Examples are $x^5+6x^4-2x^2+x-1$, $\sqrt{1+sinx}$, e^{sinx}, $xcosx$, $sinx+sin2x$. (If you can give a quick, rough sketch of *any* of these. . . .)

To be honest, there is a third class of functions you might meet:

Class 3: functions which can be graphed by first-semester methods, but only after a sequence of computations so technical and messy that you'll probably make more errors doing them than you would doing the original problem. Examples can be found in your old first-semester exam on graphing techniques.

So how do you deal with these functions that are too difficult to graph?

It's not as hard as you think. The point is, you don't have to know *everything* about the graph, you only have to have some *idea* about how things look. The idea doesn't have to be precise —and it might even be wrong! But at least it will get you thinking.

Let's consider the following problem:

$$\int_0^\pi xcosxdx = xsinx \Big|_0^\pi - \int_0^\pi sinxdx = xsinx - cosx \Big|_0^\pi =$$

$$-(-1)-(-1) = 2$$

Now I'm sure what *xcosx* looks like. (You can graph it if you want.) But I do know this: from $\pi/2$ to π, *cosx* is *negative*, and this is the interval where x is its *largest*. So the function *xcosx* is 'weighted' towards the negative. An answer of $+2$ seems too darn *positive*. There is probably a mistake.

Another example:

$$\int_0^{\pi/4}(sinx+sin2x)dx = (-cosx-2cos2x)\Big|_0^{\pi/4}$$

$$= \left(\frac{-\sqrt{2}}{2}-0\right)-(-1-2) = 3-\frac{\sqrt{2}}{2}\ .$$

Now *sinx* and *sin2x* are both *increasing* on $0<x<\pi/4$. So the function takes its largest value at the right endpoint, $f(\pi/4)=sin(\pi/4)+sin(\pi/2)=\sqrt{2}/2+1$, which is smaller than 2. The length of the interval is $\pi/4 \simeq 3.14/4$, which is smaller than 1. Therefore the integral is contained in a rectangle whose area is smaller than 2×1, and since $sinx+sin2x$ is generally much smaller than 2, we expect the integral to be substantially smaller than 2. But the calculated 'answer', $3-\sqrt{2}/2$, is actually *larger* than 2. Something smells fishy.

Sometimes you can 'bootstrap' your way up to a graph. Consider for instance the function $f(x)=xarcsinx$. This doesn't look too promising. (Set its derivative equal to zero and you get the equation $arcsinx+x/\sqrt{1-x^2}=0$. Solve *that* for x!) But we can do it by steps.

First, look at the graph of *arcsinx*:

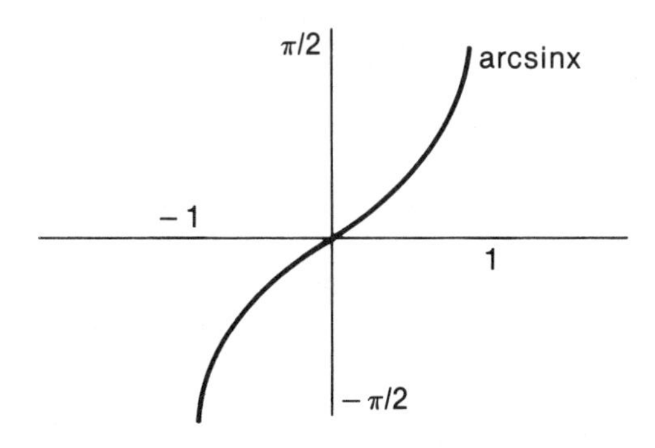

The domain of *arcsinx* is $[-1,1]$. Now in this interval, x is always smaller than 1 (obvious!). So *xarcsinx* should lie *below*

arcsinx, and moreover should connect the origin to the endpoints.

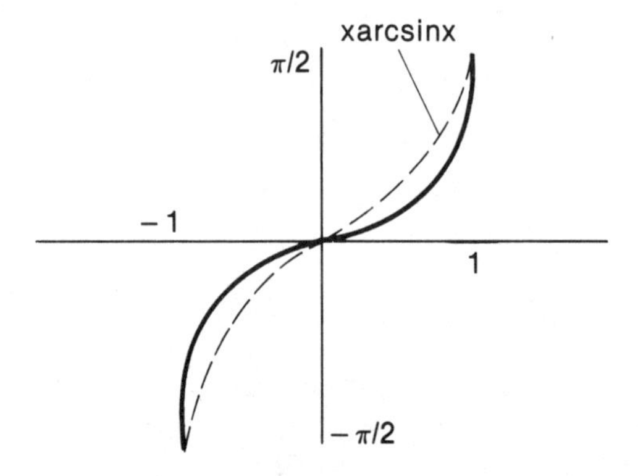

Notice one important feature: the graph I've drawn here 'flattens out' at the origin; that is to say, its derivative is zero: $f'(0)=0$. Notice one other thing: this graph is WRONG! It's OK for $x>0$, but not for $x<0$: x and *xarcsinx*, are *odd* functions, so their product, *xarcsinx*, should be an *even* function – that is, it should look the same on both sides of the y-axis.

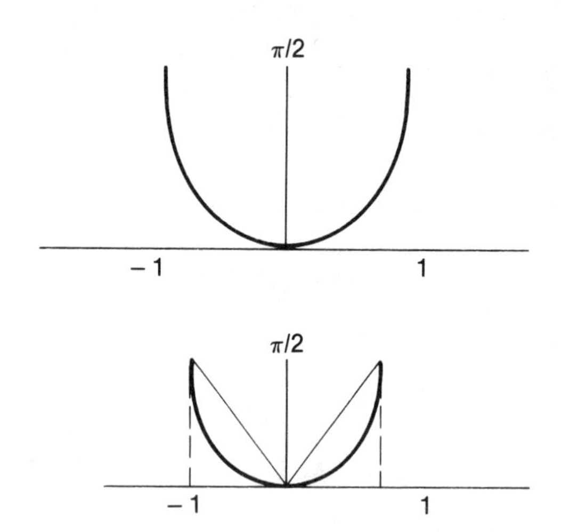

So the true graph is this one. Notice, *xarcsinx* lies below the lines connecting the origin to the endpoints, so comparing the integral to the areas of two triangles shows $\int_{-1}^{1} x\arcsin x\,dx = \pi/2$ must be wrong.

Finally, here's an interesting example:

$$\int_0^{\pi/2} cosx e^{sinx} dx = -e^{cosx} \Big|_0^{\pi/2} = -e^0 - (-e^1)$$
$$= e - 1 \simeq 1.718....$$

This answer strikes me as suspiciously large. After all, the interval is only of length $\pi/2 \simeq 1.57$, and the function starts out at $cos(0)e^{sin(0)} = 1$, and drops to $cos(\pi/2)e^{sin(\pi/2)} = 0$. Of course e^{sinx} does increase—but *cosx decreases*, so that ought to

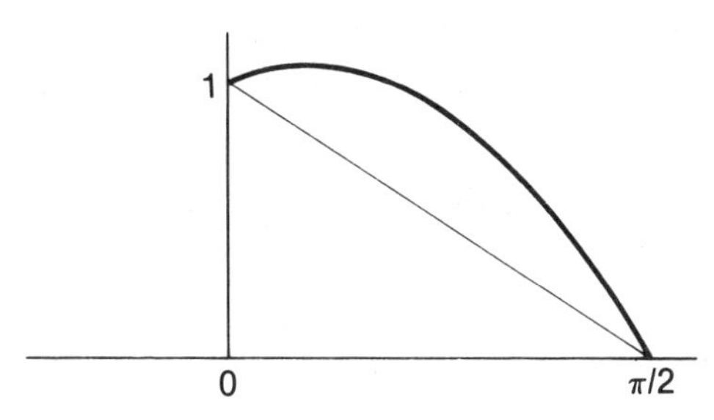

balance out. In other words, the integral should not be too much larger than the area of the triangle which connects 1 at $x = 0$ to 0 at $x = \pi/2$, and this of course has area $\frac{1}{2}(1)(\pi/2) = \pi/4 \simeq .8$. So the answer we got just seems too large.

Are you convinced?

In fact, there *is* a mistake in the solution. Here is the real, live *correct* solution:

$$\int_0^{\pi/2} cosx e^{sinx} dx = e^{sinx} \Big|_0^{\pi/2} = e^1 - e^0$$
$$= e - 1 \simeq 1.718....!!!!$$

What this means is that I *did* make a mistake—but not for the reasons I just gave! So there is an error where the error-check errs (if an error-check could check errors).

What if you'd done the problem correctly to begin with, but then believed in the error check? Well, you'd waste time pouring over your solution for a while, looking for mistakes. Then finally you might start to doubt your error check. After all, this check didn't *prove* the answer was wrong—it only made it *seem* too large. Yes, e^{sinx} *does* increase, while *cosx does* decrease—but maybe they *don't* balance out. That must be it: e^{sinx} must increase *more* than *cosx* decreases, *enough* to make the area as large as it apparently is. Yes!

So what has happened? This: we just learned something about the function $\cos x e^{\sin x}$—it increases *significantly*, before it drops to 0 at $\pi/2$. That fact may not be essential to the problem (it isn't), and you may never use it again (I'd be surprised if anyone did). Still, there is something intellectually satisfying in having discovered it. (At least *I* am intellectually satisfied: when I first cooked up this example, I intentionally wrote in

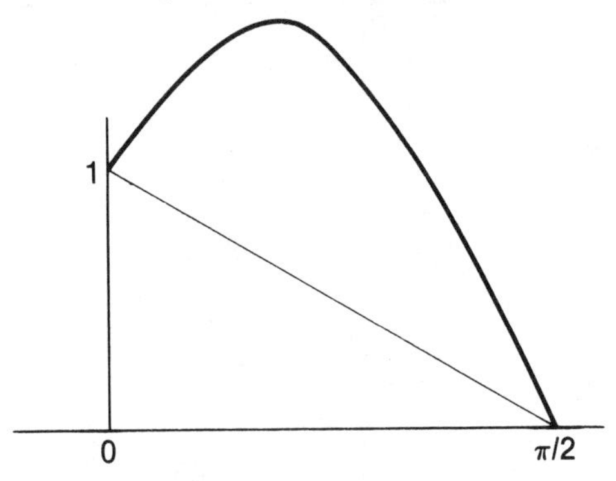

the error, and then gave the argument 'proving' the answer was wrong. I'm ashamed to admit it, but I *believed* the argument. I was astounded when I realized that the answer was indeed correct. My first reaction was to hide the whole episode, but then I decided I could use it to make a point, and appear brilliant to boot. Of course now it is perfectly obvious to me that the answer is as large as it is.)

Here are some problems where errors can be detected by comparing the graph of the function to some rectangle or triangle, or just by staring at the picture:

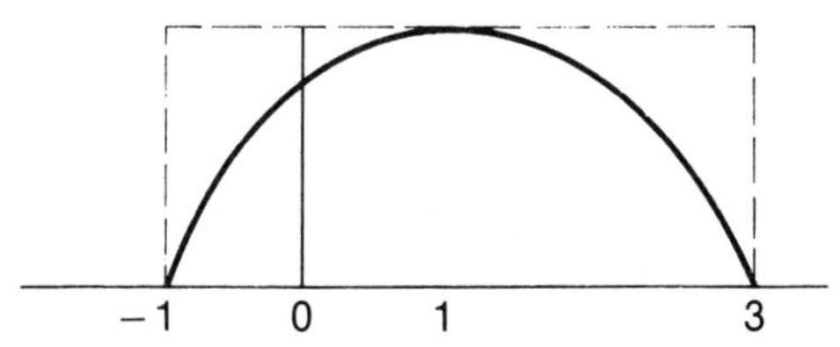

1. $\int_{-1}^{3}(1+x)(3-x)dx = \int_{-1}^{3}(3+4x-x^2)dx$

$$= 3x+2x^2-\frac{1}{3}x^3 \Big|_{-1}^{3}$$

$$= (9+18-9)-\left(-3+2+\frac{1}{3}\right) = 18\ \frac{2}{3}$$

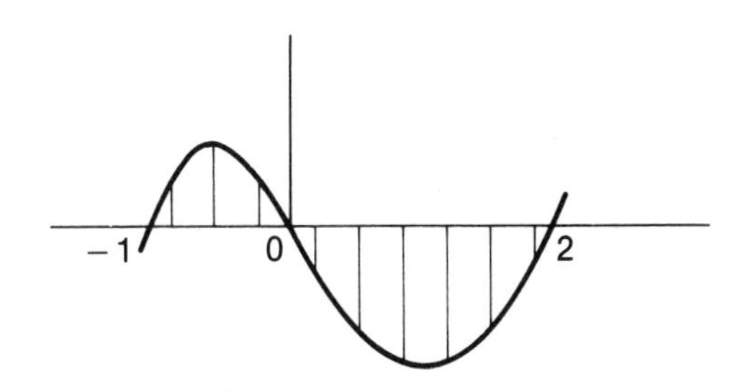

2. $\int_{-1}^{2}x(x+1)(x-2)dx = \int_{-1}^{2}x(x^2-x+2)dx$

$$= \int_{-1}^{2}(x^3-x^2+2x)dx$$

$$= \left(8-\frac{8}{3}+4\right)-\left(\frac{1}{4}-\frac{1}{3}+1\right) = 9\ \frac{3}{4}$$

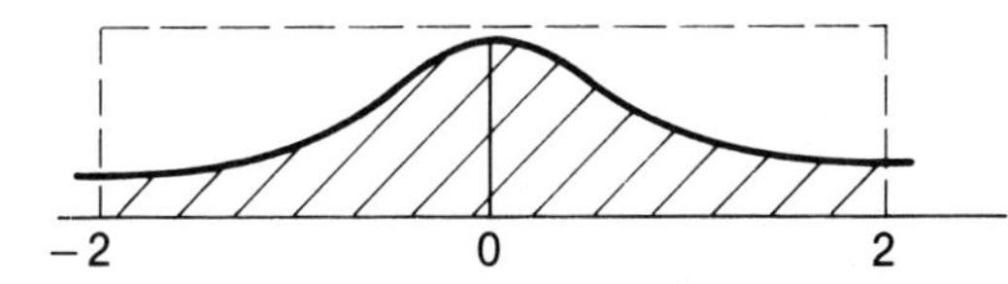

3. $\int_{-2}^{2}\frac{dx}{4+x^2} = 2arctan(x/2)\Big|_{-2}^{2} = 2\left[\frac{\pi}{4}-\left(\frac{-\pi}{4}\right)\right] = \pi$

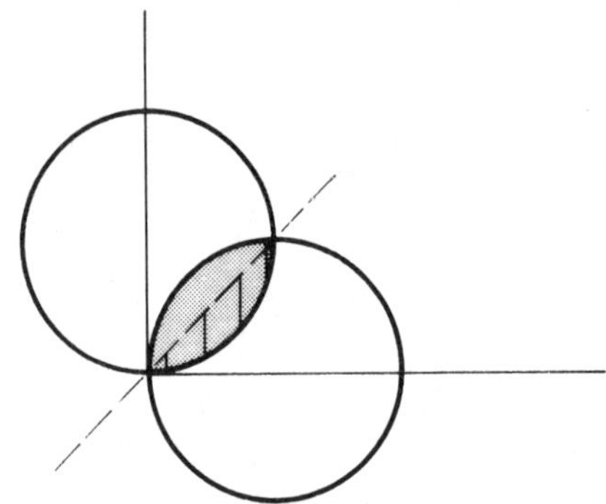

4. Suppose we want to find the dotted area in the figure above. By symmetry this area is twice the hatched area, so we have

$$A = 2\times \frac{1}{2}\int_0^{\pi/4} cos^2_\theta d_\theta = \int_0^{\pi/4}\left(\frac{1+cos2_\theta}{2}\right)d_\theta =$$

$$=\left[\frac{\theta}{2}+\frac{sin2_\theta}{4}\right]\Big|_0^{\pi/4} = \frac{\pi}{8}+\frac{1}{4}.$$

Draw pictures to show that the following definite integrals are definitely wrong:

5. $\int_0^1 arcsinxdx = 1$
6. $\int_0^1 arccosxdx = \frac{\pi}{4}$
7. $\int_0^\infty arctanxdx = \pi/2$
8. Integrating by parts with $u = arcsinx$, $dv = dx$, so that

$$du = \frac{dx}{\sqrt{1-x^2}} \quad \text{and } v = x,$$

we have

$$\int arcsinxdx = \frac{x}{\sqrt{1-x^2}} - \int \frac{x}{\sqrt{1-x^2}}dx$$

$$= \frac{x}{\sqrt{1-x^2}} + \sqrt{1-x^2} + C$$

so that $\int_0^1 arcsinxdx = \frac{x}{\sqrt{1-x^2}} + \sqrt{1-x^2}\ \Big|_0^1$, which *diverges* because of the denominator $\sqrt{1-x^2}$ at $x=1$. So the integral is improper and divergent. (Or is it? Does the picture look improper?)

4. Derivatives Again, and the Fine Art of Being Crude

Remember: a positive derivative means your function is increasing; a negative derivative means decreasing. Squeezed between these two is the fact that an "extreme point"—a maximum or a minimum—can only occur when the derivative equals zero.*

Typically (in calculus courses) you compute derivatives in order to sketch curves. The message here is that, by knowing what the curve looks like, you can tell if you differentiated wrong.

Implicit in this is the notion that you have some way of knowing what the curve looks like. How can you do that without sketching the curve? And how do you sketch curves without computing derivatives? And how do you compute derivatives without making mistakes? This is starting to look like a vicious circle.

It's not. Remember, you don't need *precise* information about a function, just general facts. Is it positive or negative? Are there any obvious zeros? Is it increasing? Decreasing? Do you expect any maximums or minimums?

For instance, the function $f(x) = x\cos x$ equals zero at $x=0$ and $x=\pi/2$ (since $\cos\pi/2=0$). Also, between 0 and $\pi/2$, both x and $\cos x$ are positive, so so is $f(x)$. We can safely conclude that $f(x)$ has a *maximum* somewhere between 0 and $\pi/2$. Thus

*At this point teachers are likely to pounce all over me: it's not strictly necessary for the derivative to equal zero at an extreme point; the derivative could simply not exist, as in $f(x) = |x|$ at $x=0$. But I have yet to see a test in which such an extreme point occurs—teachers invariably locate their max's and min's where f' equals 0. (I hope in pointing this out I am not killing the goose. . . .)

$f'(x) = cosx + sinx$ cannot be correct, because this expression is never 0 between 0 and $\pi/2$—everything about it is strictly positive (in the range where we're looking, that is).

Similarly, $f(x) = x(x-1)(x^2+1)$ has obvious zeros at $x=0$ and $x=1$. In between it's negative (because of the $x-1$), so it must have a minimum. So whatever the derivative is, it must have a zero somewhere between 0 and 1. Thus $f'(x) = 3x^2 - 2x + 1$ can't be right, since this derivative never equals zero. (The quadratic formula gives imaginary roots.) Neither is $f'(x) = 3x^2 - 2x - 1 = (3x+1)(x-1)$, since neither of its zeros ($x=1$ and $x = -1/3$) are where they should be.

Finally, at the risk of boring you completely, consider the function $f(x) = xe^{-x}$ (which we've looked at once before). If you stare at this equation for a moment, three facts stand out:

1. $f(0) = 0$
2. $f(x) > 0$ for $x > 0$
3. $f(x) \simeq 0$ for x "large"

(The third fact may not be quite so obvious to everyone, but it should be. Basically it says that "exponential decay" beats out "linear growth".) These three facts lead to three more facts:

4. $f(x)$ is increasing at $x=0$
5. $f(x)$ is decreasing for large x
6. $f(x)$ has a maximum somewhere

The interpretation into derivatives is clear:

7. $f'(0) \geq 0$
8. $f'(x) < 0$ for large x
9. $f'(\text{somewhere}) = 0$.

We can now check three candidates for the derivative:

$$f'(x) = \begin{cases} (x-1)e^{-x} \\ (x+1)e^{-x} \\ (1-x)e^{-x} \end{cases}$$

The first attempt, $(x-1)e^{-x}$, fails to satisfy conditions 7) and 8). The second attempt, $(x+1)e^{-x}$, satisfies 7), but contradicts 8) and 9). Only the third try satisfies all three conditions. This doesn't necessarily mean that it's correct (you can check it), but at least it's a *reasonable* answer.

Naturally, it's not always easy to just 'look at' an equation and tell what the function looks like. There comes a point when it's no longer worth it. I, for one, would not bother trying to figure out what something like $\sqrt{x}/e^{\sqrt{x^2+(logx)^2/4-xlogx}}$ looks like. (Go ahead and try!) But when the function is not too complicated, it's worth a try.

Geometric reasoning about slopes and derivatives is not the only way to catch mistakes. Sometimes the answer itself just doesn't make sense. A man pulling a rope in a related rates problem will not in general pull at velocities exceeding the speed of light. Nor will a rope 500 feet long ever enclose an area the size of New Mexico. Or even Rhode Island. And it certainly won't enclose a region whose area is *negative*. Take a typical 'max-min' problem for example:

> A rectangular fence is to be built. Two (opposite) sides are to be made from fancy redwood, costing $10 per foot. The other two sides are to be made from tacky chickenwire, which costs $1 per foot. If the total cost is to be $100, what is the largest area that can be enclosed by such a fence?

Our "solution" runs as follows:

Let x be the length of the redwood sides, and y be the length of the chickenwire sides. Then

$$A = xy \quad \text{(the area)}$$
$$100 = 10x+y \quad \text{(the cost)}$$

Thus

$$y = 100+10x$$
$$A = x(100+10x) = 100x+10x^2$$
$$dA/dx = 100+20x$$

Setting $dA/dx = 0$ gives

$$0 = 100+20x, \textit{ or } x = -5.$$

Stop. Something is wrong. The answer $x = -5$ can't be correct. Length *cannot* be negative. There's an error here, and we'd better take care of it. (Of course we could just fudge and let $x = +5$, but let's not.)

Perhaps you can see the mistake, but I'll tell you anyway: the equation $y=100+10x$ *should be* $y=100-10x$. If we correct this, then we have

$$\begin{aligned} A &= x(100-10x) = 100x-10x^2 \\ dA/dx &= 100-20x \\ dA/dx &= 0 \quad x = 5. \end{aligned}$$

(So fudging would have been all right!)

$$y = 100-10x = 100-50 = 50.$$

Stop again! Now what's wrong, you ask, isn't 50 a reasonable number? After all, it *is* positive. Why am I being so hypercritical?

True, $y=50$ isn't total nonsense, but it still has a problem: if two sides of a fence are each 50 feet long, that's 100 feet. Those two sides *alone* will cost \$100, leaving *nothing* to pay for the other two sides with. Calculus problems, unlike Congress, do not permit deficit spending.

It's time to make explicit a certain feature of a great many of our error checks so far: the use of *crude estimates*.

In the last example, we caught an error by *estimating* the cost of the 'solution' and realizing that it violated a condition of the problem. It was a *crude* estimate because we blatantly ignored two sides of the rectangle—they could only add more cost to something that was already straining the budget.

In earlier examples we compared the areas under curves to the areas of rectangles or triangles. In essence, we were *estimating* the areas; our estimates were *crude* because we were ignoring the curvature of the curves, we were throwing away (or tacking on) huge chunks of area.

We have also engaged in such crudities as $\pi=3.14$ and $e=2.178$. Actually, for many purposes, a 'better' approximation would be $\pi=3$ and $e=3$, not to mention $\sqrt{10}=3$, $2\sqrt{2}=3$, and even, if you really want to be crude, $2=3$ (as in $62/7=63/7=9$). How simple mathematics becomes when you estimate!

There are good reasons for being crude. When you're looking for errors, you don't want to get involved in a lot of messy computations—that would most likely just lead to more mistakes. Crude estimates *simplify* the problem, hopefully to the point where it is impossible to make any more mistakes.

Crude estimates work for a simple reason: When you makes a mistake, more often than not it's a *big* mistake. This may sound pessimistic, but it's not. Big mistakes are easy to find, it's the little ones that'll kill you. It's obvious, for instance, that a phone bill of $2240 is a bit high, while the same bill for $22.40 may or may not be overcharging you.*

One good way of making big mistakes is to multiply when you should have divided, as in

$$\frac{d}{dx}\sqrt{1+\frac{x^2}{100}} = \frac{100x}{\sqrt{1+\frac{x^2}{100}}}$$

This may look all right, but look again: from $x=1$ to $x=2$, the function $f(x)=\sqrt{1+x^2/100}$ barely budges: $f(1)=\sqrt{1.01}\simeq 1.005$, while $f(2)=\sqrt{1.04}\simeq 1.02$. The slope—that is, the derivative—can't be too large. But $100x/\sqrt{1+x^2/100}$ is *very* large between 1 and 2; in general, it's bigger than 100. That's a crude estimate: $\sqrt{1+x^2/100}\simeq 1$, so you can ignore it; the x in $100x$ is bigger than 1, so you can certainly ignore it; thus $100x/\sqrt{1+x^2/100}\simeq 100$. Now a tangent line with slope 100 is virtually indistinguishable from a *vertical* line, while our function is essentially horizontal:

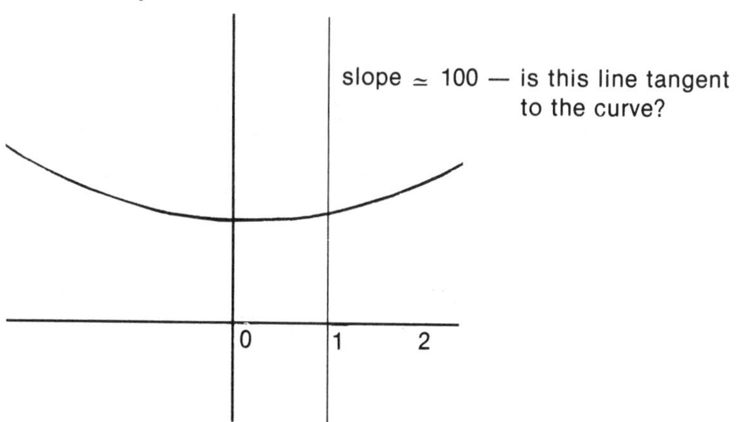

*These are 1982 dollars. Who knows what the future may bring.

Obviously there's something wrong here. (Of course, we're only off by a factor of 10,000!)

Estimates are especially useful—and easy—for word problems. Suppose we need to find the largest area possible for a right triangle of perimeter 50 inches. (Don't ask *why* we need to do this—that should be obvious.) An answer of 10,000 square inches, while at least not negative, is clearly too large: the triangle can't *possibly* be bigger than a 100×100 square. Even an answer of 1000 seems too big: the triangle will have to be considereably smaller than 50″ on a side, which puts an upper limit of $\frac{1}{2} \times 50 \times 50 = 1250$ on the area. Although 1000 is less than 1250, it is still uncomfortably close to an estimate we think is much too big.

On the other side, the answer 50 square inches is clearly too small: we could easily get by with a 10×10 triangle, which has area exactly 50. Without actually doing the computation (and I really haven't done it), I would guess that the maximizing triangle is approximately 15×15, with an area of approximately $\frac{1}{2} \times 15 \times 15 \simeq 100$.

Let me finish with a pet peeve: the approximate population of the planet. Students are routinely asked to approximate the population in 1990 if in 1970 there were 3.5 billion persons and in 1980 there are 4.1 billion. Now I don't mind an answer like 4.8 *million* or 480 billion—these are clearly way off the mark. No, what bugs me is the following answer:

$$\text{1990 population} \simeq 4{,}802{,}857{,}142.8$$

Among other things, I'd like to meet that eight-tenths of a person. Even 'rounded off' to ...143 leaves me wondering where all this accuracy comes from. (I can tell you: it comes from an overzealous pocket calculator.)

These exercises may keep you off the street for a while:

1. Let $f(x) = (x-1)(x-2)(x-3)$. Are any of the following statements at all reasonable?
   a. f is always positive
   b. f is increasing for all x
   c. f has three extreme points
   d. f is increasing at $x = 1$

e. f is increasing at $x=2$
f. f is increasing at $x=3$
g. $f'(2)=1$
h. $f'(2)=-1$
i. $f'(2)=0$
j. $f'(x)=3(x-2)^2$
k. $f'(x)=2(x-1)(x-3)$
l. $f(10)=2046$ (*estimate*, don't just compute!)
m. $f(20)=7988$

2. Find fault with these derivatives.

a. $f(x) = log(x^2-x+1) \quad f'(x) = \dfrac{1}{x^2-x+1}$

(Hint: $f(0) = f(1) = 0$.)

b. $f(x) = \sqrt{1-x} \quad f'(x) = \dfrac{x}{\sqrt{1-x^2}}$

(Hint: is this function increasing or decreasing?)

c. $f(x) = log(\sqrt{1-x^2}) \quad f'(x) = \dfrac{\sqrt{-2x}}{\sqrt{1-x^2}}$

(Hint: where does the formula for f make sense? Does the formula given for f' also make sense?) (Look at small values of x. Is $f(x)$ increasing or decreasing?)

d. $\mathrm{f}(x) = e^{1/cosx} \quad f'(x) = \dfrac{-sinx}{cos^2x}e^{1/cosx}$

3. Point out the errors in the attempts to solve the following max-min problem:
A cylinder is to be built to hold 1000 cubic inches of highly compressed pedagogical hot air. What is the smallest possible surface area (top, bottom, and shaft) for such a cylinder?

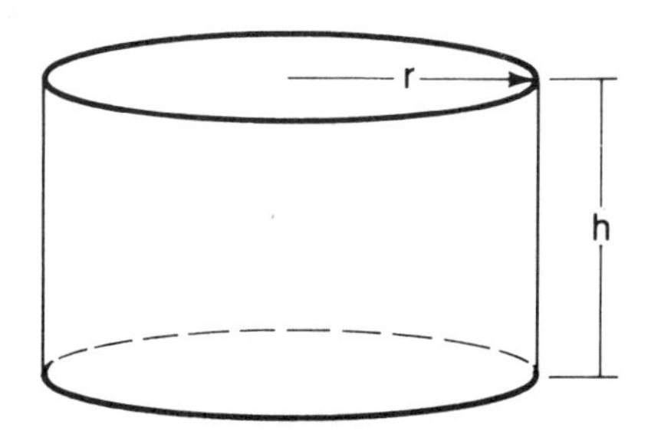

a. $V=\pi r^2h,\ A=\pi r^2+2\pi rh=\pi r^2+2/r,\ dA/dr=2\pi r-2r^2=0,$ $2r(\pi-r)=0,\ r=\pi,\ A=\pi^3+2/\pi$. (Hint: if $r=\pi$, how much area will the top alone have?)

b. $V = \pi r^2 h = 1000$, $A = 2\pi r^2 + 2\pi rh = 2\pi r^2 + 2000\pi r^3$, $dA/dr = 4\pi r + 6000\pi r^2 = 0$, $4\pi r(1 + 1500r) = 0$, $r = -1/1500$, $h = 1000\pi r^2 = 1000\pi/2250000 = \pi/2250$, $A = 2\pi/2250000 - 2\pi^2/(1500 \times 2250)$

c. $V = \pi r^2 h = 1000$, $A = 2\pi r^2 h + 2\pi rh = 2\pi r^2 + 2000/r$, $dA/dr = 4\pi r + 2000/r^2 = 0$, $r^3 = 500/\pi$, $A = 2\pi(500/\pi)^2 + 2000/(500/\pi) = 50000/\pi + 4\pi$. (Hint: how many square inches are there in the average classroom blackboard?)

d. $V = \pi r^2 h = 1000$, $A = 2\pi r^2 + 2\pi rh = 2\pi r^2 + 2000/r^2$, $dA/dr = 4\pi r + 4000/r = 0$, $4\pi r^2 = -4000$, $r = \sqrt{-1000/\pi}$; since you can't take the square root of a negative number, there is no minimum to the surface area.

4. Use crude estimates to answer the next several questions.

   a. The average high school graduate has spent how many hours in class?

      1000 5000 10000 20000 50000 1000000000000000000000000

   b. If all the books in your school library were stacked one on the other, would the stack reach the moon? the clouds? the ceiling?

   c. How much would it cost to string quarters from Maine to New Mexico?

   d. How many math problems have you done since first grade? How many have you gotten wrong? (Hint: the second estimate should not be larger than the first!)

   e. It is generally conceded that our brain cells are constantly dying *and not being replaced*. How long does it take for 10000 brain cells to die? 1 second? 1 minute? 1 hour? 1 day? 1 month? 1 year? Do you have that many brain cells?

5. The product of increasing functions is increasing. The functions $(x-1)$, $(x-2)$, and $(x-3)$ are all increasing (they each have slope $=1$). Thus $f(x) = (x-1)(x-2)(x-3) = x^3 - 8x^2 + 11x - 6$ is increasing. Thus $f'(x) = 3x^2 - 26x + 11$ is always positive. But $f'(2) = 3 \times 4 - 26 \times 2 + 11 = 12 - 56 + 11 = -23$. What is wrong here? (Keep going until you find the really big one.)

5. Reducing to Special Cases

There are few activities in life more enjoyable than making a fool out of someone else. One of the most satisfying ways of doing this is to let someone make some grandiloquent statement of high-sounding import and complexity, and then to point out a simple, obvious contradiction in what has just been said, thus undercutting the pompous jerk and causing him to tumble at your feet, where you can stomp him into the dust.

In mathematics, this is called Reducing to Special Cases, sometimes known as Plugging in Zero.

Let's begin with an example, in the form of a dialogue. PYTHAGORAS (the one in sandals) speaks to DESCARTES (the sickly youth lying on a bed cogitating).

PYTHAGORAS: I've just completed a remarkable theorem concerning the area of an ellipse. Would you care to hear it?

DESCARTES: Sure, why not?

PYTHAGORAS: The area of an ellipse is in ratio to the area of a circle as the square on the hypotenus of the right triangle whose two other sides are formed by the major and minor axes of the ellipse is in ratio to the square on the diameter of the circle.

DESCARTES: In other words, if the equation of the ellipse is $x^2/a^2+y^2/b^2=1$, then the area is $\pi(a^2+b^2)$. Is that what you mean?

PYTHAGORAS: You could put it that way.

DESCARTES: Well that's nonsense. Just take the case when a and b are equal. Then you just have a *circle* of radius a, whose area truly is πa^2, *not* $\pi(a^2 + a^2) = 2\pi a^2$ as you would have it.

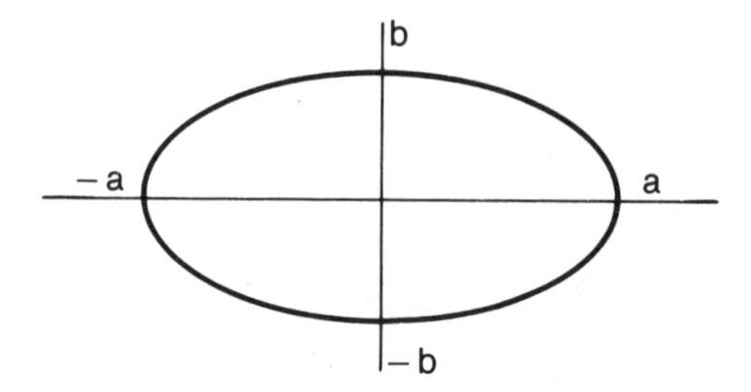

Pythagoras goes home in defeat. But, fool that he is, he doesn't give up. He reads up on analytic geometry and returns.

PYTHAGORAS: I'm back. This time I've got the right formula: $A = \pi(a^2 + b^2)/2$. See: when a and b are equal, you get πa^2, as you said it should be.

DESCARTES: That's nice. But look what happens this time when $b = 0$. The ellipse flattens out to a line, which has no area. But your formula leaves an area $\pi a^2/2$! Grovel, you degenerate!

The moral of the story is this: when confronted by a new and suspicious formula, try relating it to things which you *know* are true. Put it through the wringer. Sometimes a 'representative' example belies the formula, as did Descartes' circle (a 'typical' ellipse) at first. Other times you have to take things to their *extremes*, like Descartes' flattening out of the ellipse. The point is, if a formula is correct, it will survive these tests.

Another moral is this: formulas are ever so much better to deal with than numbers. Suppose Pythagoras came up to you and said that the ellipse $x^2/4 + y^2/9 = 1$ had area $13\pi/2$ (which comes out of his second formula). How can you argue with that number? It's wrong of course—but it doesn't miss by much! (The correct area is 6π, so we're only off by $\pi/2$, about an 8% error. it would require an accurate picture and rather subtle estimates to discover the error.)

Oftentimes when a problem is given with lots of numbers in it, it is helpful to replace those numbers with variables and solve a more 'general' problem. The advantage of having variables is that you can *vary* them. For instance, suppose we want to find the volume of a truncated cone.

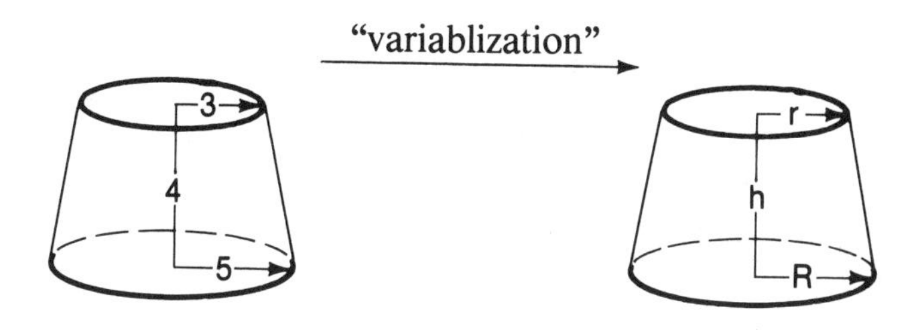

In variables,

$$V = \pi\int_0^h \left(\frac{R}{r}x\right)^2 dx = \frac{\pi R^2}{r^2}\ \frac{1}{3}x^3 \Big|_0^h$$

$$= \pi R^2 h^3/3r^2,$$

so that $V = \pi(25)(64)/3(9) \simeq 180$.

Now 180 is again not far from correct. (The correct value is approximately 200: $196\pi/3$.) But that formula is absolute garbage: when the top and bottom radii are equal (ie, $r=R$), the 'cone' becomes a cylinder, whose volume is obviously $\pi R^2 h$, in contradiction to the formula $V=\pi R^2h^3/3R^2=\pi h^3/3$. (What's worse, just consider what happens when r gets smaller and smaller: geometrically the cone is getting smaller, but the formula $V=\pi R^2h^3/3r^2$ gets *larger*! That r has no business being in the denominator!)

You occasionally meet problems where you are forced to do a certain amount of 'variabilization'. Those horrible "related rates" problems are like that. Let's consider one:

> Sand is being poured at a constant rate of 3 cubic inches per second into a pile which for no reason whatsoever always has the shape of a right circular cone whose height is half the diameter of its base. At what rate is the diameter expanding when the pile has a volume of 9π cubic inches?

Our 'solution' proceeds as follows:

$$V = \frac{\pi}{3}hr^2 = \frac{\pi}{3}r^3 \quad (\text{since } h = r = \frac{1}{2}\text{ diameter})$$

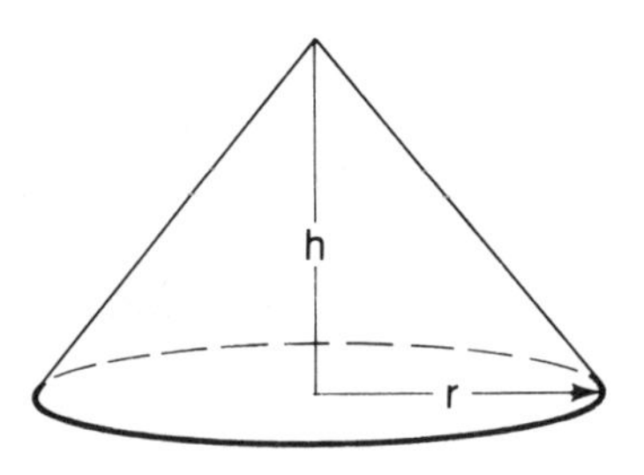

$$\frac{dV}{dt} = \pi r^2 \frac{dr}{dt} = 3$$

$$\frac{dr}{dt} = \frac{3}{\pi r^2} = \frac{3}{\pi(9\pi)^2} = \frac{3}{9\pi^3}$$

Answer: $\dfrac{1}{3\pi^3}$

This is terribly mistaken, but it's hard to see why. The problem is, the answer is a pure number, which no longer refers in any meaningful way to the problem from which it came. The most we can say is, it's got a 3 and a π in it.

Suppose, however, we went whole hog in variablizing, and assigned a variable to everything in sight. In particular, let's replace the number 3 with the symbol k, and the number 9π with the symbol V_0.

[Let me stop for a moment and say something about how you choose symbols. Usually you choose a symbol that *abbreviates* something. Thus V_0 abbreviates Volume. (Actually, I would have used V, but V is already being used, so I tacked on the subscript $_0$. This has the added advantage of suggesting that V_0 is an 'instance' of V.) The symbol k, however, came out of thin air and my own preference for that particular letter: the word "rate" abbreviates to r, which was already taken, and r_0 would suggest too strongly (much too strongly) that 3 was a particular 'instance' of the radius. In brief, use any symbol you feel like; just try to avoid confusing yourself. (When possible, though, use symbols which will confuse anyone trying to grade your paper, then complain when they grade it wrong.)]

OK, let's get on with it. If we follow our previous solution but using V_0 and k in place of 9π and 3, we get

$$V = \frac{\pi}{3} hr^2 = \frac{\pi}{3} r^3$$

$$\frac{dV}{dt} = \pi r^2 \frac{dr}{dt} = k$$

$$\frac{dr}{dt} = \frac{k}{\pi r^2} = \frac{k}{\pi V_0^2}$$

and there's the error! What in the name of Gauss are we doing replacing r with V_0? After all, r is a *distance*, while V_0 is a *volume*! (See the next section, *Dimensions*.) It becomes obvious that what we need to do is to replace r not by V_0, but by r_0, the radius corresponding to V_0. We see easily that

$$V_0 = \frac{\pi}{3} r_0^3$$

so that

$$r_0 = \left(\frac{3V_0}{\pi} \right)^{1/3}$$

Thus

$$\frac{dr}{dt} = \frac{k}{\pi r_0^2} = \frac{k}{\pi \left(\frac{3V_0}{\pi} \right)^{2/3}} \qquad (***)$$

Finally, plugging in $k=3$ and $V_0=9\pi$, we get

$$\text{Answer: } \frac{dr}{dt} = \frac{3}{\pi \left(\frac{27\pi}{\pi} \right)^{1/3}} = \frac{3}{\pi \cdot 3} = \frac{1}{\pi}$$

Part of the beauty of this is that, although I still made a mistake in the final answer (the power 2/3 became 1/3!), I would probably still get close to full credit for having written down the equation (***).

One more comment about this problem. Replacing the data 3 and 9π with symbols k and V_0 allows us to see beyond the immediate problem. For instance, from (***) we can see that, if the rate k is increased, the rate dr/dt is increased. This of course makes a lot of sense, and helps convince us that (***) is the correct formula. Is there anything interesting we can do with V_0? Yes: we can set it equal to zero! After all, when a pile of sand forms, it must start with zero volume. But what happens to (***)? It blows up! That is, when a pile of sand first begins to form, the diameter is expanding *infinitely fast*! Now I ask you: Is such a thing possible, or have I made another typical mistake?

Reducing to Special Cases and Plugging in Zero are not at all restricted to problems based on a geometric picture. They are just as effective in an algebraic setting—maybe more so, as we witness the following dialogue between CARDANO (looking secretive) and GAUSS (looking princely).

CARDANO: Give me a polynomial to factor.
GAUSS: OK: $x^{17}+1$.
CARDANO: No sooner said than done: $x^{17}+1=(x+1)(x^{16}+x^{15}+\dots+1)$.
GAUSS: Interesting: you mean that $2=1^{17}+1=(1+1)(1+1+\dots+1)=(2)(17)=34$. Very interesting. Would you like to try again?
CARDANO: No. Give me a more meaningful problem: give me a quintic.
GAUSS: Surely: x^5+x^2+2.
CARDANO: Easy: $x^5+x^2+2=(x^2+2)(x^3-2x+2)$.
GAUSS: Easier, easiest: $2=0^5+0^2+2=(0^2+2)(0^3-2\cdot 0+2)=4$.
CARDANO: Oops, I meant $x^5+x^2+2=(x^2+2)(x^3-2x+1)$. Plug zero into *that*, you upstart!
GAUSS: I'll go you one better: $4=1^5+1^2+2=(1^2+2)(1^3-2\cdot 1+1)=3\cdot 0=0$.

This conversation continues indefinitely, but it gets less educational (and more vindictive). The point is, anytime you have an equation which is supposed to hold generally (as opposed to one where you're supposed to solve for the unknowns), you can quickly check for accuracy by plugging in a few well-chosen numbers—like 0 or 1.

Of course you could have figured out that $(x^2+2)(x^3-2x+2)$ actually expands into x^5+2x^2-4x+4, rather than x^5+x^2+2—but that takes a lot of work (and are you sure I did it right?). Much more work than plugging in zero. Be lazy! Do what's easiest! Plug in zero! And if zero doesn't work, plug in one! (If one doesn't work, you're on your own.)

Here's one for you:

$$\int x arcsin x dx$$
$$= \frac{1}{2} x^2 arcsin x - \frac{1}{2} \int \frac{x^2}{\sqrt{1-x^2}} dx$$
$$= \frac{1}{2} x^2 arcsin x - \frac{1}{2} \int (\sqrt{1-x^2} - \frac{1}{\sqrt{1-x^2}}) dx$$
$$= \frac{1}{2} x^2 arcsin x - \frac{1}{2} \int \sqrt{1-x^2} dx - \frac{1}{2} \int \frac{1}{\sqrt{1-x^2}} dx$$
$$= \frac{1}{2} x^2 arcsin x - \frac{1}{2} (\frac{1}{2} x\sqrt{1-x^2} +$$
$$+ \frac{1}{2} arcsin x) - \frac{1}{2} arcsin x + C$$
$$= \frac{1}{2} [(x^2-1) arcsin x - \frac{1}{2} x\sqrt{1-x^2}] + C$$

How can we check this indefinite integral for accuracy? We can do so by making it into a *definite* integral. Remember that the *arcsine* is *positive* for $0 \le x \le 1$. Therefore $\int_0^1 x arcsin x dx$ should be *positive*. But when we evaluate, we get

$$\int_0^1 x arcsin x dx = \frac{1}{2} [(x^2-1) arcsin x -$$
$$- \frac{1}{2} x\sqrt{1-x^2}] \Big|_0^1 = 0-0 = 0.$$

Sometimes you can solve a problem by switching over to a similar, *but far simpler* problem whose solution is either obvious or at least requires very little work. Suppose, for instance, you are asked to determine whether $x=2$ is a maximum or a minimum of the function $f(x)=x^4-8x^3+16x^2-32$. (You may wish to verify that $f'(2)=0$—or are you willing to believe me for once?)

The procedure is straightforward: you just compute the second derivative, which in this case turns out to be $f''(2)=-16$ (trust me!).

OK, now what? Well, if you're like me you know that a positive second derivative means one thing, and a negative second derivative means the other—*but you can't ever remember which is which*. So what you can do is this: ignore $x^4-8x^3+16x^2-32$, and think about x^2, because x^2 is a *far simpler* problem. You should know by heart what x^2 looks like: it's a parabola, and it has a *minimum* at $x=0$. Also, you can easily get its derivatives: $2x$ and 2. Since the second derivative, 2, is *positive*, that tells you what the connection must be: a *positive* f'' means you're at a *minimum*, so a *negative* f'' must mean you're at a *maximum*. And that solves the problem: $x=2$ must be a *maximum* for $f(x)=x^4-8x^3+16x^2-32$, since $f''(2)=-16$ is negative.

Here then is the usual batch of exercises.

1. What follows are various attempts to express the volume of the truncated cone pictured in the text. Find fault with them.
   a. $V=\dfrac{1}{2}\pi(r+R)^2h$ (Hint: let $r=R$)
   b. $V=\dfrac{1}{3}\pi R^2(h+r^2/h)$ (Hint: let h 0.)
   c. $V=\pi h(R^2-r^2)$ (You're on your own now.)
   d. $V=\pi(R+r)(H+h)$
   e. $V=\dfrac{1}{3}\pi h(5Rr-R^2-r^2)$
   f. $V=\dfrac{1}{3}\pi h(R^2+Rr+r^2)$
   g. $V=\pi h(R^2-Rr+r^2)$
2. Ridicule the following 'equations':
   a. $x^6-x^5+2x^4-x^3+2x^2+1=(x^3-x^2+x+1)(x^3+x-1)$
   b. $x^4+2x^3-x+2=(x+2)(x+1)(x^2-x+1)$

c. $\sqrt{x+y} = \sqrt{x}+\sqrt{y}$

d. $log(x+y) = logx+logy$

e. $cos3\theta = 3cos\theta$

f. $cos4\theta = cos^4\theta - 3cos^3\theta + 2cos^2\theta - cos\theta + 1$

g. $cos5\theta = 16cos^5\theta - 20cos^3\theta + 5cos\theta$

3. Find fault with this argument:

$$tan2\theta = \frac{2tan\theta}{1-tan^2\theta}, \quad \text{therefore} \quad tan4\theta = \frac{4tan\theta}{1-tan^4\theta}.$$

6. Dimensions

You can't add apples and oranges.

To be sure, that's just a grade-school truism, and you might well ask, what do apples and oranges have to do with calculus and higher mathematics. As it turns out, a lot.

Mathematics can answer questions about the so-called 'real world'. (Mathematicians always put quotation marks around the words "real world"; it implies they know something the rest of us don't.) In the real world, things have dimensions: meters and miles, seconds, hours and years, ounces, pounds and kilos – not to mention ergs, joules, newtons, and all the crazy terms of electrostatics. When a problem is stated for the 'real world', the answer must be *physically realistic*. In particular, the dimensions of the answer must match up with the dimensions of what was asked for.

And you can't add apples and oranges.

Let's go back to the truncated cone. How can we disprove the following claimants for the volume?

a. $V = \frac{1}{3}\pi h^2 Rr$

b. $V = \frac{1}{2}\pi \frac{h}{R^2+r^2}$

c. $V = \pi h \frac{R^2+r^2}{R+r}$

d. $V = \frac{1}{2}\pi h(R^2+R-r+r^2)$

e. $V = \frac{1}{3}\pi h(R^2+Rr+r^2)$

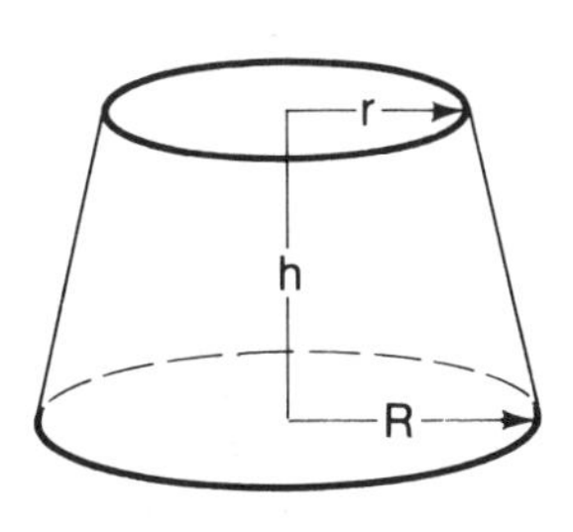

We do so by noting that all except e) violate the accepted dimensions of volumes: $length \times length \times length = (length)^3$.

The variables R, r, and h are all lengths. But then answer a) has dimensions $(length)^2 \times length \times length = (length)^4$; that can't be. Likewise for answers b) and c), which have dimensions $length/(length)^2 = (length)^{-1}$ and $length \times (length)^2/length = (length)^2 = area$, respectively.

Answer d) is even worse: adding $R^2 + R$ makes no physical sense whatsoever – you can't add area to length. You can't add apples and oranges.

Only answer e) passes the test – *at least it has the correct dimensions.* That certainly doesn't prove its correctness, but if you had to pick one answer out of the lot, it would have to be e).

Of course answers a) through d) could also be eliminated by Reducing to Special Cases. Try it, though: it requires some amount of thought (and it assumes you have a fairly good understanding of what the picture looks like, which you don't always have). You might find d) – the dimensionally ridiculous one – rather tenacious when it comes to special cases; at least it works when $R = r$!

The point is, looking at dimensions is a very quick way of checking for inconsistencies – and it gets right to the heart of many mistakes. In a long computation with lots of variables, it is very easy to lose a factor here or there, or drop an exponent. (How often have you recopied x^3 as x^2? Not as often as I have, I'll bet.) Mistakes like these stand out when you check dimensions: when you accidentally change x^3 to x^2, you're in fact changing volume into area; while there's not much difference between 3 and 2, there's a world of difference between volume and area.

Moreover, by working backwards, you can oftentimes find the *source* of the error: just keep going until you find a formula where the dimensions *do* make sense, then look for what got lost in the transition.

I personally use this kind of dimensional analysis (to give it a fancy name) a lot. Once on a physics test I got an answer to an electrostatics question that made no sense at all – I had wound up adding ohms to volts, or some such thing. So I kept going back until I found a line where at least the dimensions were correct. Then I simply filled out everything else with dielectric constants and other junk factors until *their* dimensions were all right. I had no idea what I was doing, but I was determined to have the proper dimensions.

That was the only problem I got right on that test.

So checking dimensions is good for finding mistakes in the solutions to physical problems. But what about plain old math problems?

The miracle is, it's not necessary for a problem to be stated in physical terms—you, the problem solver, can *give* it a physical interpretation.

All you have to do is to assign all the variables a dimension, just being careful not to wind up adding apples and oranges. (I.e., if your expression contains the term $a+b$, then a and b must be given the same dimension.) *Length* is always convenient—suppose we give x dimensions of length. Remember then that dx—which is just a very (very!) small amount of x—also has dimension, length. Thus

$$\frac{d}{dx}\sqrt{1+x^2/a^2} = \frac{2ax}{\sqrt{1+x^2/a^2}}$$

is just plain wrong: if x has dimensions of length, then so must the a, since x^2/a^2 is being added to 1, which is dimensionless. Now the function $\sqrt{1+x^2/a^2}$ is dimensionless, so the left-hand side has dimension *length*$^{-1}$ (remember the dx!) But the right-hand side has dimension *length*^{2}. This doesn't make sense. We have to get something out of the numerator, and the obvious thing to move is the a. Let's put it in the denominator:

$$\frac{d}{dx}\sqrt{1+x^2/a^2} = \frac{2x}{a\sqrt{1+x^2/a^2}}$$

No, that's not right either. The dimensions still don't match. This time the right-hand side is *dimensionless*. We need another a:

$$\frac{d}{dx}\sqrt{1+x^2/a^2} = \frac{2x}{a^2\sqrt{1+x^2/a^2}}$$

At last we have a realistic answer—it's still wrong (why?), but at least it makes a little sense.

Looked at from the standpoint of dimensions, the Chain Rule makes a lot of sense. For instance, $d/dx\ sin(ax+\pi) = cos(ax+\pi)$ is hopelessly wrong—I hope you can see why (remember, dx has dimensions). We need that a out front to get the thing right.*

As for integrals, the same stuff works:

$$\int (x^2+a)^2 dx = \int (x^4+2ax+a^2)\,dx = \frac{1}{5}x^5+ax^2+a^2x+C$$

is just plain wrong: if x has dimensions of length, then a must have dimension $(length)^2$—since the integral adds x^2 to a—and the answer should have dimension $(length)^5$. But ax^2 only has dimension $(length)^4$. We're adding apples and oranges!

Where did ax^2 come from? It came from integrating the $2ax$ inside the integral. But once again, $2ax$ has dimension $(length)^3$, which doesn't match up with the rest of the integral, whose dimensions are $(length)^4$. This must be where the error is.

In fact, the correction is now obvious: the only way to get from $(length)^3$ to $(length)^4$ is to change $2ax$ to $2ax^2$—and indeed, that's just what it should have been all along:

$$\int (x^2+a)^2 dx = \int (x^4+2ax+a^2)\,dx = \frac{1}{5}x^5+ax^3+ax^2+C$$

(Another way to give the $2ax$-term the correct dimensions would be to multiply in $\sqrt{a}$—but what in the world is a square-root doing in this problem? It clearly doesn't belong. (See 'The "What Did You Expect?" Method' section.))

I always have trouble remembering where junk factors go. Do they belong in the numerator, or in the denominator? For instance, which is correct?

*I put the π in this example to force a and x to have 'opposite' (i.e. inverse) dimensions. Actually, as explained later, I didn't really need the π—ax has to be dimensionless inside of $sin(\ \ )$ anyway.

$$\int \frac{dx}{x^2+a^2} = \begin{cases} a\ arctan(ax) \\ a\ arctan(x/a) \\ \frac{1}{a}\ arctan(x/a) \\ \frac{1}{a}\ arctan(ax) \end{cases}$$

I can never remember. But dimensions save the day.

If you assign a dimension to x—say, length—then you also have to assign length to the constant a. Then the integral has dimensions of $L/L^2 = length^{-1}$. (Remember, *dx also* has dimensions.) Therefore the first two answers are out: the a in front belongs in the denominator. As to the stuff inside the *arctan*, there is a simple principle: *you can't take the arctangent of anything with dimensions.* What could you possible mean by the *arctan* of 2 feet—is that the same as the *arctan* of 24 inches? How about 61 centimeters?

So you have to kill off the dimensions inside the *arctan*. Now ax doesn't do it—it gives you *area*. Only x/a works, so that must be right:

$$\int \frac{dx}{x^2+a^2} = \frac{1}{a}\ arctan(x/a) + C.$$

(Of course, another way to kill off the dimensions inside the *arctan* would be to take $arctan(a/x)$; but this can't be right, because we know a Special Case, namely $a=1$:

$$\int \frac{dx}{x^2+1} = arctan x, \text{ not } arctan(1/x).$$

The combination of Dimensions with Special Cases is a powerful one. Not only do they find a great many mistakes, but they can even be used sometimes to actually *solve* a problem where a solution is yet to be written down.)

This goes the other way also. I can remember that

$$\frac{d}{dx} arctan(x/a) = \frac{something}{1+(x/a)^2}$$

In order to remember what the 'something' is, I just think about dimensions: I need a $1/length$ (because of the dx), and it has to come from the a, which has dimensions *length*. Thus

$$something = 1/a$$

Alternatively, I might want to get that fraction x/a out of the denominator, and have

$$\frac{d}{dx}arctan(x/a) = \frac{something}{x^2+a^2}$$

This time, I'm forced into '*something*' $= a$, in order to keep things straight.

That simple principle I mentioned above is in fact a general principle for the so-called 'transcendental' functions—$sinx$, $cosx$, e^x and so on. *You can never allow a dimensioned argument inside one of these functions.*

Now this might seem mildly contradictory (as if any contradiction is ever mild). For example, the integral

$$\int sin10xcos3xdx = \frac{7}{2}cos7x - \frac{13}{2}cos\,13x + C$$

makes perfectly good sense (in spite of being incorrect), in spite of the fact that you're taking cosines of x. All this means is that x *cannot be assigned a dimension* in this problem. At least not as it stands.

But how did we get this integral? From a formula:

$$\int sinaxcosbxdx = \frac{a-b}{2}cos(a-b)x - \frac{a+b}{2}cos(a+b)x + C$$

Now in the formula we *can* use dimensions. Giving x the dimension *length*, as usual, we are forced to give a and b the dimension $length^{-1}$. Then ax and bx are dimensionless, as they must be.

But now we see that the formula is incorrect: the integral, on the one hand, has dimension *length* (because of the dx), while the right hand side has dimension $1/length$. To make good dimensional sense, we have to put the $a+b$ and $a-b$ in the denominator:

$$\int sinaxcosbxdx$$
$$= \frac{1}{2}\left(\frac{1}{a-b}\ cos(a-b)x - \frac{1}{a+b}\ cos(a+b)x\right)^*$$

At least now we have the proper dimensions. (Unfortunately, the formula is still wrong: look at what happens in the Special Case of $b=0$.) Once again, *having variables is better than having numbers.* When you have variables, you can check dimensions and you can reduce to special cases; with just numbers, you can't do either.

Here we go again: exercises.

1. Find fault with these formulas:
   a. $V = \pi R^2 h^2$
   b. $V = \pi h(R/r)^2$
   c. $A = \pi + Rr$
   d. $A = \frac{1}{2}\sqrt{(R^2-r^2)(r+r)}$
2. Where did I go wrong in the following computations?
   a. $V = \pi\int_0^R (x-r)^2 dx = \pi\int_0^R (x^2+2rx-r)dx$
   $$= \pi\left(\frac{1}{2}x^3 + rx^2 - rx\right)\Big|_0^R$$
   $$= \left(\frac{1}{3}R^3 - rR^3 + rR\right)$$
   b. (Think polar coordinates) $r^2 = acos\theta$ gives a graph whose inside area is
   $$A = \frac{1}{2}\int_0^{2\pi} a^2cos^2\theta d\theta = \frac{a^2}{2}\int_0^{2\pi}\left(\frac{1-cos2\theta}{2}\right)d\theta$$
   $$= \frac{a^2}{2}\,[\theta + sin2\theta]\Big|_0^{2\pi} = a^2\pi$$
3. Which derivatives make dimensional sense?
   a. $$\frac{d}{dx}\left(\frac{x^2+ax+b}{(x-c)(x-d)}\right) = \frac{(x-c)(2x+a)}{(x-d)^2} + \frac{(x-d)(2x+a)}{(x-c)^2} - $$
   $$- \frac{x^2+ax+b}{(x-c)^2(x-d)^2}$$

*The typesetter just lost one point for leaving off the $+C$.

b. $\dfrac{d}{dx} cos(ax^2+b) = -(2ax+b)sin(ax^2+b)$

c. $\dfrac{d}{dx} log(ax) = \dfrac{1}{ax}$

d. $\dfrac{d}{dx} log(ax) = \dfrac{a}{x}$

4. Pick out the dimensionally correct answer:

$$\int \frac{dx}{a^2x^2+b^2} = \begin{cases} \frac{1}{ab} arctan(ax/b) \\ \frac{a}{b} arctan(ax/b) \\ \frac{b}{a} arctan(x/ab) \\ \frac{1}{ab} arctan(x/ab) \\ \frac{x}{b} arctan(abx) \end{cases}$$

5. By special cases and dimensions, pick the right answer:

$$\int e^{-x^2}dx = 2 \Rightarrow \int e^{-x^2}dx = \begin{cases} \sqrt{2\pi a} \\ \sqrt{2\pi/a} \\ \sqrt{2\pi}/a \\ \sqrt{2\pi}a \end{cases}$$

6. Does the (true!) formula $log(ab) = log(a) + log(b)$ make dimensional sense?

7. Symmetry (The Same Thing Over and Over)

Symmetry is a concept dear to the hearts of mathematicians. Inherent in nature, first appropriated by the classical geometers and artists, it now finds itself expressed in all kinds of mathematics, including the formulas of algebra and calculus. To a mathematician, there is nothing like the thrill of looking at a mathematical formula and seeing in it all the balance, harmony and sheer beauty that the subject is capable of.

Be that as it may, symmetry is also a good, quick way of catching lots of mistakes.

Basically, a problem or a formula shows symmetry when certain parts of it are *interchangeable*. That is, symmetry is involved anytime something looks the same from two ostensibly different viewpoints. Since examples are better than abstraction, here are a few.

The formula $A = 1/2hb$ (area $= 1/2$ height $\times$ base) shows symmetry: height and base are on an equal footing; if you interchange them—if you tip the triangle over—the formula doesn't change. A formula such as $A = 1/2h(h+b)$ is *not* symmetric—h and b play different roles; if you tipped this triangle over, it would suddenly have a different area.

Likewise, the equation $(x+y)^2 = x^2 + 2xy + y^2$ stands a good chance of being correct, because both sides are symmetric: interchanging x and y does not change either side. On the other hand, $(x+y)^3 = x^3 + 2x^2y + 3xy^2 + y^3$ is clearly *wrong*: x and y are interchangeable on the left-hand side, but not on the right. Similarly, $(x-y)^3 = x^3 - 3x^2y - 3xy^2 + y^3$ is wrong, this time because the right-hand side is unchanged under the exchange of x and y, while the left-hand side picks up a minus sign.

The function $f(x)=(x+a)(x+b)(x+c)$ is symmetric in the constants a, b, and c. If you try differentiating it and get $f'(x) = 3x^2+(a+b+c)x+a(b+c)$, you can see immediately where at least one mistake is: the last coefficient, $a(b+c)$, is no longer symmetric. In the integral, $\int f(x)dx = 1/4x^4 + a/3x^3 + b/2x^2 + cx$, *none* of the coefficients is symmetric.* (Why, for instance, should the a go with the x^3, rather than the b or c?)

This cuts the other way, also. The function $f(x)=(ax+b)/(2x-1)$ is *not* symmetric in a and b, so it should be suspicious when $f'(x)=ab/(2x-1)^2$ *is* symmetric.

An especially nice example of symmetry is the surprising formula for the area of a triangle (*any* triangle) whose three sides are of length A, B, and C. Suppose six students derive six formulas:

a. $\frac{1}{2}(A+B)C$

b. $\frac{1}{4}\sqrt{(A^2+B^2-C^2)(A+B+C)}$

c. $\frac{1}{2}(A+B+C)(A-B+C)$

d. $\frac{1}{2}(A-B+C)(A+B-C)$

e. $\frac{1}{4}\sqrt{(A+B+C)(A+B-C)(A+C-B)(B+C-A)}$

f. $\frac{1}{4}\sqrt{(A+B)(A+C)BC}$

Now at least one of these answers—b)—can be thrown out for improper Dimensions, and several more could probably be eliminated by judicious Special Cases. (For example, b) could be thrown out again because it gives area $=0$ in the case of a right triangle.) But Symmetry gets at the answer immediately, and without any calculation: Answers a) and b) are thrown out because in them C gets treated differently from A and B. Answer c) is junked because B gets treated differently, while A is the reason for deciding that d) and f) can't be correct. Only

*Except the first: ¼ remains ¼, no matter what you do to a, b, and c.

answer e) treats A, B, and C as equals. The answer must be e) (Is it really?)

The symmetry of answer e) stands out even more clearly if we introduce a new variable, $S = \frac{1}{2}(A+B+C)$. Then

$$\text{Area} = \sqrt{S(S-A)(S-B)(S-C)}$$

A, B, and C are now very obviously interchangeable.

In general, symmetry is useful when for some reason—geometric, algebraic, or aesthetic—you expect your answer to show some kind of symmetry, and it doesn't. The method is quick—you can frequently see at a glance that something's out of whack. It also gives you a good idea of *where* the mistake is: in $(x+y)^3 = x^3 + 2x^2y + 3xy^2 + y^3$, the mistake is not in the x^3 or y^3, but in the $2x^2y + 3xy^2$.

One more thing bears repeating: variables are ever so much better than numbers. You can do things with variables that you could never think of doing to plain old numbers. Besides, numbers sometimes show phony symmetries. For instance, the expression 124×421 has a nice, balanced symmetry to it, but the actual product, 52204, shows no symmetry or balance whatsoever.

When possible, therefore, replace numbers with variables. As an example, let's reconsider that max-min problem about the fence made from two sides wood and two sides chickenwire. (After all, we never did solve the thing.) Recall that wood cost \$10/foot, chickenwire cost \$1/foot, we had \$100 to work with, and we wanted to enclose as much area as possible.

Variablize: Let $A=10$, $B=1$, and $C=100$. Let x be the length of a side made from wood, and y be the chickenwire-side length. Then $\text{Area} = xy$, and $C = Ax + By$.

Start solving: $y = C-(A/B)x$, $\text{Area} = x(C-(A/B)x) = Cx - (A/B)x^2$; differentiating gives $C - 2(A/B)x = 0$, so $x = BC/2A$, $y = C-(A/B)x = C - (A/B)\,BC/2A = C/2$, and the 'answer' is

$$\text{Area} = \frac{BC^2}{4A}.$$

This answer, I tell you, is clearly wrong, because it violates the symmetry between A and B. What symmetry? you may ask. Aren't A and B *necessarily* different? Isn't $A=10$ and $B=1$?

No, not really, By replacing 10 and 1 with A and B, we have changed the problem by *generalizing* it. The problem now reads as follows: A fence is to be made from two materials, costing \$$A$/foot and \$$B$/foot. If the total cost is to be \$$C$, what is the largest area attainable? Stated this way, the interchangeability of A and B is obvious, and the lack of symmetry in the solution $BC^2/4A$ is apparent.

(If you're still unconvinced, use a Special Case: suppose the chickenwire were *free*, i.e. let $B=0$. Could the maximum area then really be 0?)

Not only does symmetry help point out mistakes, it can also help to solve problems by cutting down the amount of computation you have to do—and that helps to cut down the number of mistakes you'll make. A good example is partial differentiation of a symmetric function:*

$$\frac{\partial}{\partial x}\sqrt{x^2+y^2+z^2} = \frac{2x}{\sqrt{x^2+y^2+z^2}}$$

Now it is senseless to repeat for $\partial/\partial y$ the thought processes that went into figuring out $\partial/\partial x$. Instead, one simply replaces x by y:

$$\frac{\partial}{\partial y}\sqrt{x^2+y^2+z^2} = \frac{2y}{\sqrt{x^2+y^2+z^2}}$$

The same goes for z:

$$\frac{\partial}{\partial z}\sqrt{x^2+y^2+z^2} = \frac{2z}{\sqrt{x^2+y^2+z^2}}$$

(There is one drawback to this: if you do the first part wrong—as I have—then everything else will be wrong too. However, it's just as likely that you would repeat the same error over and over again anyway if you just redid the computation each

*Technically, this goes beyond what you get in first-year calculus. But the example is simple enough that this shouldn't matter.

time, so you might as well use symmetry and save some time for working on other problems.)

One nice use of symmetry is to avoid excess minus signs, which always cause trouble. Instead of

$$\int_{-2}^{2}(x^4-2x^2+1)dx = \frac{1}{5}x^5 - \frac{1}{3}x^3 + x \Big|_{-2}^{2}$$

$$= \frac{32}{5} - \frac{8}{3} + 2$$

$$- \frac{32}{5} - \frac{8}{3} - 2 = \frac{-16}{3}$$

you can use the symmetry of even functions: $x^4 - 2x^2 + 1$ looks the same on each side of zero, so the integral from -2 to 2 will just be *twice* the integral from 0 to 2:

$$\int_{-2}^{2}(x^4-2x^2+1)dx = 2 \times \int_{0}^{2}(x^4-2x^2+1)dx$$

$$= 2\left(\frac{1}{5}x^5 - \frac{1}{3}x^3 + x\right) \Big|_{0}^{2} = 2\left(\frac{32}{5} - \frac{8}{3} + 2\right) = \frac{152}{15}$$

(Notice especially that we didn't even bother with the lower limit—it obviously gives zero.) For *odd* functions, it's even easier:

$$\int_{-2}^{2}(x^3-5x)dx = 0$$

requires no computation at all, just the realization that the two sides of 0 cancel each other out. Even a mixture of even and odd is do-able—just separate the even from the odd:

$$\int_{-1}^{1}(x^4+4x^3+x^2-3x-5)dx$$
$$= \int_{-1}^{1}(x^4+x^2-5)dx + \int_{-1}^{1}(4x^3-3x)dx$$
$$= 2\int_{0}^{1}(x^4+x^2-5)dx + 0$$
$$= 2\left(\frac{1}{5} + \frac{1}{3} - 5\right) = \frac{-144}{15} .$$

The triple combination of Special Cases, Dimensions, and Symmetry is a powerful one. A correct answer must endure probes from all three directions; a wrong answer will almost always crack under such duress. For instance, that answer to the rectangular fence, Area $= BC^2/4A$, failed to show an appropriate symmetry (between A and B). Nor did it satisfy the special case $B = 0$. (The area should be infinite when one of the materials is free.) For that matter it doesn't have the correct dimensions either: remember, C was measured in dollars (yes, dollars is a dimension!), while A and B were measured in dollars/feet. Thus $BC^2/4A$ has dimensions *dollars*^{2}, instead of *feet*^{2}.

Not only do Special Cases, Dimensions, and Symmetry confirm a correct answer and expose a wrong one, they can also be used on occasion to *produce* a correct answer, seemingly out of nowhere. Since we've just about beaten it into the ground anyway, we might as well actually solve that fence problem, but without a bunch of computation.

First, a Special Case: if $A = B$, then neither side costs more than the other, so by Symmetry we should have equal sides:

$$x = y = C/4\mathrm{A}$$

(Note: $C/4A$ has the correct dimensions, *feet*. Also, as C increases—that is, as more money becomes available to build a bigger fence—the sides increase, which is sensible. Likewise, as A decreases—i.e. as materials get cheaper—the sides again get longer; in particular, if $A = 0$, then the material is free, and we can build an infinitely large fence.) Thus when $A = B$,

$$\text{Area} = \frac{C^2}{16A^2}$$

(The dimensions still check.) We now want to get the B into the equation. We might be tempted, by Symmetry alone, to write

$$\text{Area} = \frac{1}{2}\left(\frac{C^2}{16A^2} + \frac{C^2}{16B^2}\right)$$

but a Special Case proves this wrong: As B gets larger ($B \to \infty$), the area should get smaller and vanish (the rectangle can have no width), but the formula above leaves a residual area of $C^2/32A^2$. We see that what we need is actually

$$\text{Area} = \frac{C^2}{16AB} \quad \text{or} \quad \frac{C^2}{8(A^2+B^2)}$$

or, more generally,

$$\text{Area} = \frac{C^2}{\alpha AB + \beta(A^2+B^2)}$$

where α and β have to be figured out. (Anything else, such as $C^2/(8A^2+8AB)$, is clearly repulsive on grounds of Symmetry.) To figure out α and β we use two Special Cases: when $A=B$, the denominator should be $16A^2$, so we have $\alpha A^2 + \beta(A^2+A^2) = (\alpha+\beta)A^2 \Rightarrow \alpha+\beta=16$. On the other hand, if $A=0$, the area will be infinite (because one of the materials is so cheap), so the denominator should be zero:

$$\alpha(0) + \beta(0+B^2) = \beta B^2 \Rightarrow \beta = 0$$

Thus $\alpha+\beta=16 \Rightarrow \alpha=16$, so the true, final, correct answer is

$$\text{Area} = \frac{C^2}{16AB},$$

a beautiful, symmetric, properly dimensioned answer that at least makes sense when $A=B$, $A=0$, and $A=\infty$.

Practice looking for symmetries and the lack thereof in these exercises.

1. $\int sin ax sin bx dx = \frac{1}{a+b} sin ax cos bx$
2. $(ax+b)^3 = ax^3 + abx^2 + ab^2x + b^3$
   (Hint: interchange a and x.)
3. $f(x) = (a^2+b^2+c^2-x^2)^{1/2} \quad f'(a) = \frac{c-ba}{(b^2+c^2)^{1/2}}$
4. $f(x) = e^{ax+b} + e^{bx+a} \quad f'(x) = (a+b)e^{ax+b}$
5. $\frac{1}{(x-a)(x-b)} = \frac{b}{x-a} - \frac{a}{x-b}$
6. $tan(\theta+\phi) = \frac{tan\theta + tan\phi}{1-tan^2\theta}$
7. $\int \sqrt{1+abx}\, dx = \frac{1}{a}(1+abx)^{3/2} + C$

8. $f(x) = \left(\dfrac{abx}{a+b+x}\right)$ $f'(x) = \dfrac{a^2b}{(a+b+x)^2}$

9. $\dfrac{\partial}{\partial x}\left(\dfrac{xyz}{x+y+z}\right) = \dfrac{y^2z}{(x+y+z)^2}$ (Hint: see Problem #8)

10. If $\dfrac{\partial}{\partial x}\left(\dfrac{xyz}{x+y+z}\right) = \dfrac{y^2z+yz^2}{(x+y+z)^2}$, what are $\dfrac{\partial}{\partial y}$ and $\dfrac{\partial}{\partial z}$ of the same?

11. Use symmetry to simplify or solve these integrals:
    a. $\int_{-\pi}^{\pi} x \sin x dx$
    b. $\int_{-\pi}^{\pi} x \cos x dx$
    c. $\int_{-4}^{4}(x^2-x+1)dx$
    d. $\int_{-1}^{2} x^3 dx$ (Hint: $\int_{-1}^{2} = \int_{-1}^{1} + \int_{1}^{2}$)

12. The formula $\sqrt{S(S-A)(S-B)(S-C)}$ for the area of a triangle is not symmetric if we interchange, for instance, A and S. Is that a problem?

13. Point out all the gaps and faulty logic in how we finally 'solved' that stupid fence problem.

14. Use the same (flawed and faulty) reasoning by Special Cases, Dimensions, and Symmetry to get at formulas for the following:
    a. The area of an ellipse, with the equation $\dfrac{x^2}{a^2} + \dfrac{y^2}{b^2} = 1.$
    b. The volume of an ellipsoid with the equation $\dfrac{x^2}{a^2} + \dfrac{y^2}{b^2} + \dfrac{z^2}{c^2} = 1.$
    c. The volume of a truncated cone, of height h, upper radius r, and lower radius R. (Hint: when $r=0$, the Special Case formula is $\pi R^2 h/3$. Take it from there.)

    (General Hint: Think circles, spheres, and cylinders.)

8. The "What Did You Expect?" Method

How many times have you fretted over the resubstitution of x's and $\sqrt{1-x^2}$'s for *cos*'s and *sec*'s in a trig-substitution integral, only to realize when it's all done that you're just back with the same square root as you started with?

Well what did you expect?

The essence of error-checking is to use some simple feature of the problem to predict some simple feature of the answer—and then seeing if your answer has that simple feature. For instance, in the first section, positivity of a function predicted positivity of its definite integral. In some problems the feature is even more apparent: it often happens that some *piece* of the problem must itself reappear as some piece of the answer. In particular, square roots of polynomials don't just go away. Nor do they mutate into new polynomials. Similarly, exponential functions are almost impossible to get rid of—you have to beat them to death with a log.

In the same vein, certain things can occur in an answer only if they already were part of the problem: trigonometric functions do not spontaneously generate, nor do the denominators of fractions. This also obviously holds for the exponentials and square roots already mentioned.

We're going to use this to our advantage.

Let $f(x)=\sqrt{1+x^2}$. If we differentiate, we get $f'(x)= x/\sqrt{1+x^2}$. In particular, we have the same square root that we started with. In fact, you can *never* get rid of a square root (or any kind of root, square, cube or otherwise) by differentiating.

Thus

$$f(x) = x\sqrt{1-2x^2} \text{ implies } f'(x) = \sqrt{1-2x^2} - \frac{4x^2}{\sqrt{1-2x^2}}$$

$$= \frac{1-2x^2-4x^2}{1-2x^2} = \frac{1-6x^2}{1-2x^2}$$

is clearly wrong because the derivative has no square root.

Similarly, integration rarely kills off a square root—with two major exceptions:

$$\int \frac{dx}{\sqrt{1-x^2}} = arcsinx + C$$

$$\int \frac{dx}{x\sqrt{x^2-1}} = arcsecx + C$$

The upshot of this is that, presented with an integral involving a square root, you should expect the answer to involve that same square root, or else an arcsine or arcsecant. And if you don't get one, you'd better know why.

$$\int \sqrt{x} logx dx = 4\int w^2 logw dw$$

(change of variables, letting $x = w^2$

$$= 4\left[\frac{1}{3}\, w^2 logw - \int \frac{1}{3}\frac{w^2}{w}\, dw\right]$$

(integrating by parts)

$$= \frac{4}{3}\left[w^2 logw - \frac{1}{2}\, w^2\right] = \frac{2}{3}\,(xlogx - x) + C$$

cannot be correct: what happened to the $\sqrt{x}$? Likewise

$$\int \frac{dx}{\sqrt{x^2+1}} = \int \frac{sec^2\theta\, d\theta}{\sqrt{tan^2\theta+1}} = \int sec\theta d\theta = log\,\left|\, tan\theta + sec\theta \,\right|$$

$$= log\,\left|\, x + \sqrt{x^2-1}\,\right| + C$$

is nonsense. The answer does have a square root in it, but it's the *wrong* one! $\sqrt{x^2+1}$ has vanished; $\sqrt{x^2-1}$ is a poor substitute. (To be really convinced of the incorrectness of this solution, just notice that the original integral makes sense for $-1 \leq x \leq 1$, but the 'answer' does not.)

Exponentials are also forever. You can't differentiate them away, nor integrate them away. Only logarithms can kill exponentials. (Zero and infinity also zap e^x out of existence—but they are just special values of the log-function: $0 = log(1)$, while $\infty = -log(0)$.) Consequently any problem that starts with an exponential function had better *end* with an exponential function—and the same one, at that. Thus

$$f(x) = e^{3x^2+5x} \text{ gives } f'(x) = (6x+5)e^{6x+5}$$

is utterly ridiculous, as is

$$\int e^{2x}dx = e^{x^2} + C.$$

Of course these are rather simpleminded examples. (No one would ever make such mistakes, right? Wrong!) But how about the following integral:

$$\int \frac{1}{x^3}\, e^{-1/x}dx$$

A simple change of variables (which unfortunately we wrote in a miserable, unreadable scrawl) changes this integral into $\int u^3 e^u du/u^2 = \int ue^u du = (u-1)e^u + C$ (integrating by parts for the last equality). Now what was that stupid change of variables? Well, since we started with $e^{-1/x}$, we should wind up with $e^{-1/x}$. So that's what u must be: $u = -1/x$. What did you expect? Thus we get the 'answer'

$$\int \frac{1}{x^3}\, e^{-1/x}dx = \left(\frac{-1}{x} - 1\right)e^{-1/x} + C.$$

(Unfortunately, this answer is also wrong: if you stick in definite limits, like 1 and ∞, you get

$$\int_1^\infty \frac{1}{x^3}\, e^{-1/x}dx = \left(\frac{-1}{x} - 1\right)e^{-1/x}\Big|_1^\infty$$

$$= (-1)e^{-0} - (-2)e^{-1} = \frac{2}{e} - 1$$

which is *negative*, whereas the original function is *positive*.

Can you find the missing minus sign?)

There is also a certain permanence to trigonometric functions.

$$\int \frac{dx}{\sqrt{1-sin^2 x}} = arcsin(sinx) = x + C$$

seems weird: how do you differentiate x and wind up with $1/\sqrt{1-sin^2x}$?

The next example is a somewhat subtler—perhaps borderline—variant of the "What Did You Expect?" method. I'm sticking it in here partly because I couldn't find anywhere else to put it.

$$\begin{aligned}\int cos^4\theta sin^6\theta d\theta &= \int \left(\frac{1+cos2\theta}{2}\right)^4 \left(\frac{1-cos2\theta}{2}\right)^6 d\theta \\ &= \int \frac{(1+cos2\theta)^2(1-cos2\theta)}{32}\, d\theta \\ &= \frac{1}{32}\int (1+2cos2\theta+cos^2 2\theta)(1-cos2\theta)d\theta \\ &= \frac{1}{32}\int (1-cos2\theta-cos^2 2\theta-cos^3 2\theta)d\theta\end{aligned}$$

and we're not going any further with this mess, because it's already obviously wrong.

It's obviously wrong for simple (if not obvious) reasons: the function we started out with had total sine-cosine exponent of $4+6=10$. The half-angle trick, when you think about it, has the effect of cutting that exponent in *half*. (Think about it!) So we should *expect* to get a highest power of *5*, whereas all we *did* get was a highest power of *3*. Somehow we've lost two powers of $cos2\theta$. Where did they go?

Finally, a word about those nasty critters, fractions.

I claim the following integral is wrong:

$$\int_0^1 (x^4+3x^3+2x-1)dx = \frac{19}{70}$$

How do I know? Aside from the fact that I was the one who

made up the problem, I know because of the 70 in the answer: 70 has a factor of 7. *Where did that 7 come from?*

When you're dealing with polynomials, denominators arise in only two ways: as products of other denominators already present somewhere in the problem, or by integrating the polynomial: x integrates to $\frac{1}{2}x^2$, x^2 to $\frac{1}{3}x^3$, etc. In the polynomial above, neither of these processes ever produces a 7. So where does the 7 come from? It can only come from a mistake.

(The factor 10 of 70, on the other hand, is OK—the integral of x^4 gives a 5 to the denominator, while $3x^3$ gives a 4; so in fact the denominator 20 is permissible. In fact, that's where the 70 came from: a sloppily written 20.)

So numerators may do strange things, but there must be a *reason* for getting the denominators that you get.

This cuts the other way also:

$$\int_0^1 (x^6 + 6x^3 - 4)dx = -8$$

is wrong because it *doesn't* have a 7 in its denominator. You see, $\int x^6 dx$ gives a denominator 7, and 7 *never again* appears in any other denominator. *You cannot get rid of a denominator that only appears once.* Try it.

(On the other hand, watch out:

$$\int_1^4 x^2 dx = \frac{1}{3} x^3 \Big|_1^4 = 22$$

is correct (check me!), in spite of lacking a 3 in the denominator. (What denominator?) Why? Because it really does appear *twice*: once in the upper limit, and once in the lower limit. Curiously enough, though, $4 - 1 = 3$.)

Here, as always are some exercises. What did you expect?

Point out absurdities in the following derivatives and integrals.

1. $f(x) = e^{arcsinx}$ $f'(x) = \dfrac{e^x}{\sqrt{1-x^2}}$
2. $f(x) = (1-x^2)^{-1/2}$ $f'(x) = -2x(1+x^2)^{-3}$
3. $f(x) = cos(5x^2)$ $f'(x) = -10xsin(10x)$

4. $\int \frac{dx}{\sqrt{1+x^2/2}} = \frac{1}{2} \int \frac{dw}{\sqrt{1+w^2}} = \frac{1}{2}\, log(w+\sqrt{1+w^2})$

$= \frac{1}{2}\, log(w+\sqrt{1+2x^2})+C$

5. $\int e^{2x} dx = e^{x^2}+C$

6. $\int \frac{dx}{x\sqrt{4+x^2}} = \frac{-1}{4} log\left|\frac{1+\sqrt{1+x^2}}{x}\right|+C$

7. $\int_0^1 (x^7+1)dx = \frac{100}{99}$

8. $\int_0^1 (x^{99}+1)dx = \frac{100}{99}$

9. $\int_0^1 (x^{99}+1)dx = \pi+e$

9. Some Common Errors

Most of us are prone to making certain mistakes, or certain kinds of mistakes, over and over again. Some people always mix up the minus sign when they differentiate *sinx* and *cosx*. Others multiply when they should divide, as in $\int x^2dx = 3x^3 + C$. Personally, I tend to make mistakes in addition, such as in adding up students' test scores: $88 + 72 + 81 + 83 = 314$, 320 being the cut-off for a B.*

These are the so-called "stupid mistakes" everyone complains about making. It would be nice if there were some surefire way of dealing with them.

Unfortunately, the handiest hint I can think of for remedying such errors is Socrates' maxim: Know thyself. If you know you're going to make a mistake, you may not be able to avoid it, but at least you can catch yourself right away, and (hopefully) correct it. So go ahead and write down $d/dx \cos x = \sin x$, but, if you know you tend to get this wrong, ask yourself, Did I get it right this time? You can always stick in the minus sign, and no one'll be the wiser.

The mistakes you make will be as unique as your fingerprints or your handwriting (unless you're copying from someone else's paper). Nevertheless, there are errors that are common enough that, perhaps by pointing them out here, we can take steps toward their control. What follows then, though by no means an exhaustive survey of the common errors of calculus, is at least an introduction. I have chosen six categories of

*Recently I've taken to using a pocket calculator, but I still make mistakes —now I push the wrong buttons.

common mistakes: 1) missing minus signs; 2) disappearing parentheses; 3) lost coefficients; 4) dropped or otherwise damaged exponents; 5) fractional inversion (sounds pretty forbidding, doesn't it?); 6) uncontrollable computations.

1. Missing minus signs.

Aside from the controversy over what to do with *sinx* and *cosx*, there is always the Chain Rule to be dealt with, and the problems of denominators. As far as I know, there is still a reward out for the correct differentiation of

$$\frac{1}{1-(1-x^{-2})^{-3}}$$

(There's a bonus for integrating it as well). The best thing to do in a situation like this is to call for help. If help is unavailable (or recalcitrant), I suggest *counting* the minus signs, being sure to include one for the denominator. Alternatively, you can try determining if the function is increasing or decreasing. This probably won't help, but it's better than nothing.

Also, it goes without saying (but we'll repeat it anyhow): area, volume, and stuff like that are never negative.

2. Disappearing parentheses.

Parentheses are a way of keeping straight what goes with what. When you leave them off you run the risk of doing your test score serious harm. For instance,

$$\int \frac{2xdx}{(x^2+1)^2} = -1/x^2+1+C$$

looks all right, if you remember that what you really mean is $-1/(x^2+1)+C$, but most likely you'll eventually convert it to $(-1/x^2)+1+C$ (at which point the C should really scoop up that 1, but nevermind). This gives strange answers: the integral $\int_0^1 2xdx/(x^2+1)^2$ looks perfectly well-behaved, yet

$$-1/x^2+1 \Big|_0^1$$

blows up at the lower limit. There seems to be an infinite amount of area beneath this unassuming curve!

Parentheses also have a tendency to disappear in differentiating:

$$\frac{d}{dx}(x^2+x-1)^4 = 4(x^2+x-1)2x+1$$

Actually, disappearing parentheses is a problem that may itself soon disappear. Most students who leave out parentheses do so because they don't fully understand what the parentheses are there for. But as students become more accustomed to working with computers, where oftentimes a program won't even run if you don't stick in enough parentheses, they will (hopefully) be impressed early on with the necessity of stating things precisely. [The computer age in general may tend to make this book obsolete, but I doubt it (at least I hope not!). The human potential for error is boundless, it's something we can always count on. Computers may eventually take all our derivatives and do all our integrals (there are already languages which do this), but we'll still be setting up the problems and pushing the buttons, and we'll keep on doing those things wrong. Computers allow us to handle bigger and more complicated problems; our mistakes will likewise get bigger and more complicated. In fact, as we remove ourselves further and further from the computational drudgery of mathematics, letting machines handle all that, it becomes increasingly important to ask the question, What does this answer mean? Can this answer be correct? The computer won't be able to answer that; all it can say is what it's always said: Garbage in, garbage out. We are ultimately responsible for our own mistakes.]

3. Lost coefficients.

This happens when you differentiate—

$$\frac{d}{dx}(x^4+5x^3-x+1) = 4x^3+3x^2-1,$$

when you integrate—

$$\int (x^4+5x^3-x+1)dx = \frac{1}{5}x^5+\frac{1}{4}x^4-\frac{1}{2}x^2+x+C,$$

or simply when you recopy a line—

$$x^4+x^3-x+1.$$

(In case you missed it, the coefficient 5 has been ignored, as if it were never there.) Lost coefficients can be hard to detect. If it's a large enough number, or if it's a funny number like π or e, the 'What Did You Expect?' method is helpful. (Whatever happened to that $972\pi^{47}$, anyway?) If the coefficient is a variable, as in $d/dx(ax+7)^3=3a(x+7)^2$, checking for dimensions can identify the problem. It's the small constants, 2, 3, and 4 (not to mention -1), that cause the most trouble. You either have to check over your work very carefully (which never seems to work), or wait until you get nonsense (negative area, etc.) for a final answer, which then obligates you to go back and dig up the mistake, wherever it is. (This assumes the problem has some eventual meaning to it, which not every calculus problem does, especially on tests. Also, it occasionally happens—let's admit it—that you actually did the problem correctly, in which case you'll never be able to find the mistake.)

4. Dropped exponents.

This error is frequently seen in company with the preceding mistake. As you copy and recopy a formula, something like this may happen:

$$\begin{aligned} x^5-4x^4+3x^2-x+1 &= x^5-4x^4+3x^3-x^2+1 \\ &= x^5+4x^4+x^3-x+1 \\ &= x^5+4x^3+x^2-x+1 \\ &= 1+x-x^2+4x^3+x^4 \\ &= \mathit{etc.} \end{aligned}$$

Fractional and negative exponents can suffer an even worse fate:

$$\frac{1+(1+x^2)^{-3/2}}{x^{5/2}-1} = \frac{1+(1+x^2)^{3/2}}{x^{1/2}-1} = \frac{1+(1+x)^3/2}{x^{1/2}-1}$$

$$= \left(1+\frac{(1+x)^3}{2}\right)(x^{1/2}-1)^{-1}$$

$$= (1+\left(\frac{1+x}{2}\right)^3)(x^2-1)^{\underset{-}{9}1}$$ (that's the printer's idea of a smudge.)

$$= (1+\left(\frac{1+x}{2}\right))^3(x^2-1)^4$$

and so forth. One might wonder whatever happened to the square roots, or the power of 5 (in the $x^{5/2}$)? If you can assign dimensions to things (which is hard to do here, since x has to be dimensionless in order to get added to 1), you're in good shape. Otherwise, reread the section on lost coefficients.

5. Fractional inversion.

Fractions are the last straw in a great many people's mathematical training. People who would rush into a burning building to save a child, who confidently make decisions to buy this stock or that, who write informed articles on the global political situation, are all too often reduced to fear and trembling (not to mention loathing) when faced by a math problem involving 'things in the denominator'.

Even those of use who made it past fractions still have our problems with them. What does it mean to divide one fraction by another? Or into another? (And is there a difference?) And even if we 'understand' all this, we still make mistakes:

$$\int \frac{1}{300}\, t^3 dt = \frac{1}{75}\, t^4 + C$$

Reason: when you integrate t^3 to t^4, you *divide* by 4; that's just what we did—300 divided by 4 is 75.

It's very common in calculus for students (and teachers as well, though presumably not quite as much) to multiply when they should have divided, or divide when they should have multiplied:

$$\frac{d}{dx}\left(1+\frac{x}{10}\right)^{1/3} = \frac{10}{3}\left(1+\frac{x}{10}\right)^{-2/3}$$

$$\frac{d}{dx}(1+10x)^{1/3} = \frac{1}{30}(1+10x)^{-2/3}$$

$$\int\left(1+\frac{x}{10}\right)^{1/3}dx = \frac{4}{30}\left(1+\frac{x}{10}\right)^{-2/3}+C$$

This happens especially when you're first learning to integrate. You're used to differentiating, where you multiply, so you keep doing that:

$$\int sinaxdx = acosax + C.$$

(Later, having done nothing but integrals for what seems like forever, you realize that you can no longer differentiate correctly.) And then there are fractional powers: people who would never write $\int x^3dx = 4x^4 + C$ will invariably write $\int x^{1/3}dx = \frac{4}{3}x^{4/3}$.

I really don't have any suggestions here. Dimensions help, when you can assign them. Beyond that, you're on your own. Good luck.

6. Uncontrollable computations.

Teachers may be cruel, but they are not perverse.* Whatever else this may mean, it certainly means the following: On a one-hour calculus exam, you are not supposed to wind up doing fifty-nine minutes of computational arithmetic. Thus if you're asked to differentiate $(x^2+5)^{13}$, you are *not* supposed to start expanding the polynomial. Even a problem such as to differentiate $(x^4-1)(x^2+1)^2$ is best done as $4x^3(x^2+1)^2+2x(x^4-1)(x^2+1)$, and left at that—why should you expand things out if

*I may have that backwards—teachers might be perverse rather than cruel—but the conclusions are the same.

the teacher didn't? (Furthermore, what'll happen to you if you make a mistake in the expansion?)

For some problems a certain amount of computation is unavoidable, but even then you shouldn't let it get out of hand. By way of example, let's start with the innocuous test problem

$$\int \frac{(x+1)(x+5)}{(x-1)(x+2)(x+3)}\,dx$$

Of course this integral must be done by that absolute misery, partial fractions:

$$\frac{(x+1)(x+5)}{(x-1)(x+2)(x+3)} = \frac{A}{x-1} + \frac{B}{x+2} + \frac{C}{x+3}$$

so

$$(x+1)(x+5) = A(x+2)(x+3) + B(x-1)(x+3) + C(x-1)(x+2),$$

or

$$x^2+5x+5 = A(x^2+5x+6) + B(x^2+2x-3) + C(x^2-x+2),$$

which gives us three equations in three unknowns:

$$\begin{aligned} A+B+C &= 1 \\ 5A+2B-C &= 5 \\ 6A-3B+2C &= 5 \end{aligned}$$

From here on out, I'll just show the steps as they might appear on a test paper, without explanation, rhyme or reason—see if you can figure out what I've been thinking:

$$\begin{aligned} 6B-\ 2C &= 1 \\ 28B-17C &= -5 \end{aligned}$$

$$\begin{aligned} 102B-34C &= 17 \\ 56B-34C &= -10 \\ \hline 158B \qquad &= 7 \end{aligned} \qquad \boxed{B = 7/158}$$

$$6\left(\frac{7}{158}\right) - 2C = 1$$

$$\frac{42}{158} - 2C = 1$$

$$-2C = 1 - \frac{45}{158} = \frac{158-45}{158} = \frac{113}{158}$$

$$\boxed{C = \frac{-113}{316}}$$

$$A + \frac{7}{158} - \frac{113}{316} = 1$$

$$A = 1 - \frac{7}{158} + \frac{113}{316}$$

$$= \frac{158 \times 316 - 7 \times 316 + 158 \times 113}{158 \times 316}$$

$$= \frac{49928 - 2212 + 17854}{49928}$$

$$\boxed{A = \frac{62570}{49982}}$$

$$\begin{array}{r} {\scriptstyle 3\,2} \\ 158 \\ \underline{316} \\ 948 \\ 158 \\ \underline{474} \\ 49928 \\ \underline{17854} \\ 67782 \\ \underline{2212} \\ 62570 \end{array} \qquad \begin{array}{r} {\scriptstyle 1\,2} \\ 158 \\ \underline{113} \\ 474 \\ 158 \\ \underline{158} \\ 17854 \end{array}$$

It's obvious there's an error here somewhere: no teacher would ever put a problem on a test whose answer involves so much computation and such large numbers. (Perhaps I should amend that: no *reasonable, humane* teacher would do such a thing. Your teacher may be different.) This is supposed to be a problem on a *calculus* exam, not an arithmetic test. I should expect the problem to be computationally easy, not messy; I should have been suspicious as soon as I got a denominator 158 for B. As it is, notice I never did finish the problem – I never integrated anything. Why? *Because I ran out of time trying to do all those stupid multiplications!**

Of course in 'real life' (those quotation marks again!), the problems you are handed are not artificially constructed so as

*Set a timer for five minutes, and *do* the integral that I never finished.

to be computationally easy. One must distinguish between the classroom and reality. But even so, anytime you find yourself doing an inordinate amount of arithmetic—or any other kind of unpleasant work—you should stop for a moment and ask yourself if what you're doing is really necessary, what's the point of it, is the problem really this hard, isn't there some easier way? You may find some surprising answers.

Only one exercise: Look into your own soul—and your old math papers, if you haven't destroyed them—and ask yourself, What kind of fool am I?

Suggested Reading

G. Polya, *How To Solve It*
Perhaps the best book around on problem-solving.

Raymond Weeks, *Boy's Own Arithmetic*
An entertaining collection of problems, worth perusing.

E.A. Maxwell, *Fallacies in Mathematics*
You can prove almost anything if you make the right mistakes.

J. Hadamard, *The Psychology of Invention in the Mathematical Field*
Interesting stuff, at least to a mathematician.

Morris Kline, *Why the Professor Can't Teach*
No comment.